METHODS IN MOLECULAR BIOLOGY

Series Editor
John M. Walker
School of Life Sciences
University of Hertfordshire
Hatfield, Hertfordshire, AL10 9AB, UK

For further volumes:
http://www.springer.com/series/7651

Environmental Microbiology

Methods and Protocols

Second Edition

Edited by

Ian T. Paulsen

Department of Chemistry and Biomolecular Sciences, Macquarie University, Sydney, NSW, Australia

Andrew J. Holmes

School of Molecular Bioscience, University of Sydney, Sydney, NSW, Australia

Humana Press

Editors
Ian T. Paulsen
Department of Chemistry
and Biomolecular Sciences
Macquarie University
Sydney, NSW, Australia

Andrew J. Holmes
School of Molecular Bioscience
University of Sydney
Sydney, NSW, Australia

ISSN 1064-3745 ISSN 1940-6029 (electronic)
ISBN 978-1-4939-5903-7 ISBN 978-1-62703-712-9 (eBook)
DOI 10.1007/978-1-62703-712-9
Springer New York Heidelberg Dordrecht London

Softcover reprint of the hardcover 2nd edition 2014

Printed on acid-free paper

Humana Press is a brand of Springer
Springer is part of Springer Science+Business Media (www.springer.com)

Preface

This new volume on Environmental Microbiology provides an up-to-date view of molecular mechanisms for investigating microbial communities and their biological activities. In particular, it looks at recent advances that have a big impact on the field such as metagenomics and other "omics" technologies, NanoSIMS, and stable isotope probing. As such, this volume should be of broad general interest not only to scientists working directly in environmental microbiology, molecular microbiology, and genomics but also to industrial scientists and educators in molecular microbiology.

This volume is organized into four sections: the first looks at methods involved in sampling environmental microorganisms, the second profiles different methods for investigating the diversity and composition of microbial communities, the third focuses on techniques for analyzing biological activities in situ, and the final section examines high-throughput "omics" approaches for the characterization of environmental microbial communities.

It is an exciting time for environmental microbiologists, and some of the technical advances outlined in this volume should provide an unprecedented glimpse into the structure, composition, and activity of microbial communities across diverse environments and illuminate their impact on global ecological processes. We sincerely thank all of the contributors for sharing their technical knowledge with the wider environmental microbiology community.

Sydney, NSW, Australia

Ian T. Paulsen
Andrew J. Holmes

Contents

Part III Investigation of Biological Activities In Situ

Part IV High Throughput Genomic Approaches for Environmental Microbiology

Contributors

GUY C.J. ABELL • *CSIRO Marine and Atmospheric Research and Wealth from Ocean Flagship, Hobart, TAS, Australia*

JILLIAN F. BANFIELD • *Department of Earth and Planetary Science, Policy, and Management, University of California, Berkeley, CA, USA; Department of Environmental Science, Policy, and Management, University of California, Berkeley, CA, USA*

DAN G. BOARDMAN • *Kroto Research Institute, Sheffield University, Sheffield, UK*

LEVENTE BODROSSY • *CSIRO Marine and Atmospheric Research and Wealth from Ocean Flagship, Hobart, TAS, Australia; Department of Bioresources, Austrian Institute of Technology, Seibersdorf, Austria*

LAUREN BRAGG • *Advanced Water Management Centre, The University of Queensland, St. Lucia, QLD, Australia; Australian Centre for Ecogenomics, The University of Queensland, St. Lucia, QLD, Australia; CSIRO Mathematics, Informatics and Statistics, St. Lucia, QLD, Australia*

KATY BREAKWELL • *Department of Chemistry and Biomolecular Sciences, Macquarie University, Sydney, NSW, Australia*

ANTHONY J. BRZOSKA • *School of Biological Sciences, University of Sydney, Sydney, NSW, Australia*

EUNICE C. CHERN • *National Risk Management Research Laboratory, Office of Research and Development, U.S. Environmental Protection Agency, Cincinnati, OH, USA*

YI VEE CHEW • *School of Molecular Bioscience, University of Sydney, Sydney, NSW, Australia*

JOHN B. CLIFF • *The Centre for Microscopy, Characterisation, and Analysis, The University of Western Australia, Crawley, WA, Australia*

PAUL A. DAVISON • *Kroto Research Institute, Sheffield University, Sheffield, UK*

LIAM D.H. ELBOURNE • *Department of Chemistry and Biomolecular Sciences, Macquarie University, Sydney, NSW, Australia*

SCOTT GIFFORD • *Department of Marine Sciences, University of Georgia, Athens, GA, USA*

MICHAEL R. GILLINGS • *Department of Biological Sciences, Macquarie University, Sydney, NSW, Australia*

KARL A. HASSAN • *Department of Chemistry and Biomolecular Sciences, Macquarie University, Sydney, NSW, Australia*

RICHARD A. HAUGLAND • *National Risk Management Research Laboratory, Office of Research and Development, U.S. Environmental Protection Agency, Cincinnati, OH, USA*

ROSS HOLLAND • *National Oceanography Centre, Southampton, UK*

ANDREW J. HOLMES • *School of Molecular Bioscience, University of Sydney, Sydney, NSW, Australia*

WEI E. HUANG • *Kroto Research Institute, Sheffield University, Sheffield, UK*

LIISA KAUTTO • *Department of Chemistry and Biomolecular Sciences, Macquarie University, Sydney, NSW, Australia*

MENGQIU LI • *Kroto Research Institute, Sheffield University, Sheffield, UK*

AMANDA M. MACKIE • *Department of Chemistry and Biomolecular Sciences, Macquarie University, Sydney, NSW, Australia*
SOPHIE MAZARD • *Department of Chemistry and Biomolecular Sciences, Macquarie University, Sydney, NSW, Australia*
MARY ANN MORAN • *Department of Marine Sciences, University of Georgia, Athens, GA, USA*
HELENA NEVALAINEN • *Department of Chemistry and Biomolecular Sciences, Macquarie University, Sydney, NSW, Australia*
CATHERINE A. OSBORNE • *Department of Civil Engineering, Centre for Water Sensitive Cities, Monash University, Clayton, VIC, Australia*
MARTIN OSTROWSKI • *Department of Chemistry and Biomolecular Sciences, Macquarie University, Sydney, NSW, Australia*
CHONGLE PAN • *Chemical Sciences Division, Oak Ridge National Laboratory, Oak Ridge, TN, USA*
YAO PAN • *Department of Bioresources, Austrian Institute of Technology, Seibersdorf, Austria*
IAN T. PAULSEN • *Department of Chemistry and Biomolecular Sciences, Macquarie University, Sydney, NSW, Australia*
LINDSAY PEED • *National Risk Management Research Laboratory, Office of Research and Development, U.S. Environmental Protection Agency, Cincinnati, OH, USA*
BRANDON SATINSKY • *Department of Marine Sciences, University of Georgia, Athens, GA, USA*
DAVID J. SCANLAN • *School of Life Sciences, University of Warwick, Coventry, UK*
HENDRIK SCHÄFER • *School of Life Sciences, University of Warwick, Coventry, UK*
ACHIM SCHMALENBERGER • *Department of Life Sciences, University of Limerick, Limerick, Ireland*
ORIN C. SHANKS • *National Risk Management Research Laboratory, Office of Research and Development, U.S. Environmental Protection Agency, Cincinnati, OH, USA*
MANO SIVAGANESAN • *National Risk Management Research Laboratory, Office of Research and Development, U.S. Environmental Protection Agency, Cincinnati, OH, USA*
NANCY STRALIS-PAVESE • *Department of Bioresources, Austrian Institute of Technology, Seibersdorf, Austria; Department of Biotechnology, University of Natural Resources and Applied Life Sciences, Vienna, Austria*
JUNIOR TE'O • *Department of Chemistry and Biomolecular Sciences, Macquarie University, Sydney, NSW, Australia*
CHRISTOPH C. TEBBE • *Institut für Biodiversität, Bundesinstitut für Ländliche Räume, Johann Heinrich von Thünen-Institut, Braunschweig, Germany*
SASHA G. TETU • *Department of Chemistry and Biomolecular Sciences, Macquarie University, Sydney, NSW, Australia*
CLAIRE L. THOMPSON • *Department of Biogeochemistry, Max-Planck-Institute for Terrestrial Microbiology, Marburg, Germany*
GENE W. TYSON • *Advanced Water Management Centre, The University of Queensland, St. Lucia, QLD, Australia; Australian Centre for Ecogenomics, The University of Queensland, St. Lucia, QLD, Australia*
YUN WANG • *Kroto Research Institute, Sheffield University, Sheffield, UK*
ANDREW WARD • *Central Laser Facility, Research Complex at Harwell, Science and Technology Facilities Council, Rutherford Appleton Laboratory, Harwell Oxford, Didcot, Oxfordshire, UK*
DAYI ZHANG • *Kroto Research Institute, Sheffield University, Sheffield, UK*
MIKHAIL V. ZUBKOV • *National Oceanography Centre, Southampton, UK*

Part I

Recovery and Investigation of Biological Samples

Chapter 1

Methods for Isolation and Cultivation of Filamentous Fungi

Helena Nevalainen, Liisa Kautto, and Junior Te'o

Abstract

Filamentous fungi are important organisms for basic discovery, industry, and human health. Their natural growth environments are extremely variable, a fact reflected by the numerous methods developed for their isolation and cultivation. Fungal culture in the laboratory is usually carried out on agar plates, shake flasks, and bench top fermenters starting with an inoculum that typically features fungal spores. Here we discuss the most popular methods for the isolation and cultivation of filamentous fungi for various purposes with the emphasis on enzyme production and molecular microbiology.

Key words Filamentous fungi, Plate cultures, Liquid cultures, Fermentation

1 Introduction

Filamentous fungi are a diverse group of eukaryotic organisms with one common feature, that is, their nutrition. Fungi are heterotrophic (chemo-organo-heterotrophs) in nature which means that they are not capable of photosynthesis and thus require organic matter for growth and energy formation [1]. Fungi can live as saprophytes on dead plants and animals or their wastes or parasites assimilating tissues of living plants and animals. A typical fungal life cycle features formation of threadlike vegetative hyphae which form a mycelium, a three-dimensional structure of hyphae capable of effective assimilation of nutrients and aggressive growth. Hyphae emerge from germinating spores (conidia) that may be uni- or multinucleate, haploid or diploid. Fungi are typically isolated by plating a sample (e.g., soil, organic matter, liquids) on a Petri dish containing a rich medium such as malt extract agar and potato dextrose agar (PDA) supporting the growth of a variety of fungi. In addition to the nutrients available, the main external factors affecting the fungal growth include pH, temperature, humidity, and light. The type and concentration of carbon and nitrogen source and the cultivation temperature are amongst the most important physical factors having an effect on the type of reproduction

Ian T. Paulsen and Andrew J. Holmes (eds.), *Environmental Microbiology: Methods and Protocols*, Methods in Molecular Biology, vol. 1096, DOI 10.1007/978-1-62703-712-9_1, © Springer Science+Business Media, LLC 2014

(e.g., vegetative vs. sexual) in fungi that possess these life cycles [1]. Sporulation can be induced by the selection of the growth medium, humidity of the cultivation environment, and, in some cases, light. It should be noted that especially in the soil environment, fungal species capable of aggressive sporulation are easily overrepresented in the samples; therefore, dilution of the sample before plating is recommended to expose the less abundant species.

Fungi can be grown in liquid cultures for various purposes such as enrichment of a type of fungus of interest, production of fungal biomass, and production of enzymes and antibiotics [2]. For industrial purposes, filamentous fungi are grown in fermenters (up to 100,000 L in volume) where the cultivation parameters can be controlled and the process automated. The so-called solid-state fermentation can be performed in vessels especially designed for this type of culture. One typical application of solid culture is production of fungal mycelia and spores to be applied for biological control [3]. Solid culture is also seen as a way to modify the enzyme profiles produced by fungi as the profiles may differ from those produced in liquid culture. An excellent overview of the principles of solid-state fermentation is provided in [4].

In this chapter we describe the most popular media and methods for the isolation and maintenance of filamentous fungi on plate cultures, growing hyphae for the isolation of genomic DNA and RNA, and screening fungal colonies for enzyme activity. We also discuss setting up of various types of shake cultures and growing fungi in a laboratory fermenter.

2 Materials

2.1 Plate Cultures

Plate cultures usually contain agar as a solidifying agent. Some agars, such as PDA, are enriched with nutrients and can be purchased as "ready-made," whereas minimal agars require addition of relevant nutrients such as a carbon and nitrogen source. In addition to nutrients essential for growth, various other components can be introduced into the agar media. For example, antibiotics may be added for the selection of fungal transformants or to prohibit bacterial contamination. The surfactant Triton-X100 is typically applied to restrict the growth of fungal colonies on the plates. Here we describe the preparation of PDA plates and some variations of their use and minimal agar plates with an example of screening of xylanase activity secreted by fungal colonies.

2.1.1 Components for PDA Plates

1. PDA (e.g., Difco #213400, Voigt Global Distribution Inc, Lawrence, KS, USA; Oxoid # CM0139, Oxoid Australia Pty Ltd, Adelaide, SA, Australia).
2. Triton-X100 (e.g., Amresco, Inc., Solon, OH, USA, #0694), 10 % (v/v) stock prepared in purified water.

3. Hygromycin B (e.g., Calbiochem #400049, 1MU, Calbiochem-Novabiochem Pty Ltd, Alexandria, NSW, Australia).
4. Cellophane discs cut to fit into the 9 cm Petri dishes (*see* **Note 1**).
5. Whatman filter paper discs (#3001-917) cut to the size of the cellophane discs.
6. Glass 9 cm Petri dish to autoclave and store the cellophane discs.

2.1.2 Components for Minimal Agar Plates

1. Minimal agar (Oxoid Agar, Technical no. 3, #LP0013).
2. KH_2PO_4, $(NH4)_2\ SO_4$, MgSO4, $CaCl_2$, 5 M KOH (e.g., from Sigma Chemical Company, St Louis, MO, USA).
3. 100× mineral stock (100 mg $FeSO_4 \times 7H_2O$; 20 mg $MnSO_4 \times 4H_2O$; 20 mg $ZnSO_4 \times 7H_2O$; 40 mg $CoSO_4 \times 7H_2O$ to 200 mL of distilled H_2O).
4. Birch wood xylan 0.5 % (w/v) (e.g., Sigma #X-0502, *see* **Note 2**).
5. NaCl 1 M solution made in purified water.
6. Congo Red 1 % (w/v) (e.g., Sigma #C6767).

2.1.3 Glass- and Plasticware

1. Sterile plastic or glass Petri dishes, vented. Standard size for fungal cultures is 9 cm in diameter and 1.3 cm or 2.0 cm in height (*see* **Note 3**).
2. Graded glass containers for aliquoting and autoclaving the solubilized agar with magnetic stirrers placed in the beakers for efficient mixing of the contents.
3. Glass pipettes (10 mL) and measuring cylinders.

2.1.4 Other Equipment

1. Autoclave (bench top, free standing, industrial, etc.).
2. Pressure cooker (optional).
3. Incubator cabinets set at 28 °C to grow the plate cultures.

2.1.5 Components for Preparing the Inocula

1. Solution of 0.9 % (w/v) NaCl + 0.01 % (v/v) Tween 80, as in Subheading 3.1.4. The solution may be aliquoted into test tubes ready for use (10 mL/tube) and tubes autoclaved.
2. Well-sporulating fungal cultures on PDA plates or spore suspension stored in the cryogenic storage solution (12.4 % (v/v) glycerol and 0.04 % (v/v) Tween 80) at −80 °C.
3. Automatic pipettes plus sterile pipette tips (100 and 1,000 μL).
4. Small glass funnels plugged with permeable cotton wool, wrapped in foil and autoclaved using the dry program option.
5. Sterile glass or plastic spreads ("hockey sticks").
6. Sterile flat-slanted wooden toothpicks or glass/plastic inoculation rods.
7. Sterile 15 mL test tubes and a Bunsen burner.

2.2 Liquid Cultures

Liquid cultures are typically carried out in conical (Erlenmeyer) flasks placed on a shaker. This type of culturing is usually performed with a view of testing particular properties of the fungal strains of interest such as production of an enzyme or a metabolite. Fungi are excellent protein secretors; thus, high amounts of proteins can be found in the culture supernatants. Composition of the growth medium and cultivation conditions depend on the goal of the experiment. The procedure involves choosing the carbon or the nitrogen source and setting the pH and the shaker speed. A fungal shake culture in a laboratory is typically carried out in 50 mL of medium placed in 250 mL Erlenmeyer flasks, but cultivations can also be carried out on a smaller or a larger scale such as 15 mL test tubes, 11 mL "Duetz System" deepwell plates [5], and laboratory fermenters ranging from 0.5 to 20 L. Here we describe a growth medium and cultivation conditions that are suitable for the production of a large number of hydrolytic enzymes including cellulases and xylanases relevant for enzymatic hydrolysis of lignocellulosic biomass to fermentable sugars for biofuel production.

2.2.1 Components for Hydrolase-Inducing Medium

1. Avicel cellulose (Avicel® PH-101, Fluka #11365, from Sigma Chemical Company-Aldrich, MO, USA), soybean flour, Type I: not roasted (e.g., from Sigma #S9633), lactose, KH_2PO_4, $(NH4)_2SO_4$, $MgSO_4$, $CaCl_2$, 5 M KOH, Tween 80 (optional).

2.2.2 Components for Standard Shake Cultures

1. 250 mL Erlenmeyer flasks and suitable caps. Plastic or metal caps are recommended (*see* **Note 4**).
2. Graded glass containers and measuring cylinders or large-volume pipettors for aliquoting the medium into the shake flasks.
3. Glass pipettes, automatic pipettes, and sterile pipette tips.
4. Orbital shaker with suitable clamps to hold the conical flasks safely at the shaking speed of 250 rpm.
5. Hemocytometer (e.g., Neubauer # 717810, BLAUBRAND®, BRAND GMBH + CO KG, Germany) for spore counting. Spores will be counted from the suspension prepared in 0.9 % NaCl (w/v) + 0.01 % Tween 80 on a PDA as in Subheading 3.1.4 and filtered through a sterile cotton funnel.

2.2.3 The Duetz System

1. Cover clamp which holds microtiter plates that can fit onto an orbital shaker (# CR1700, BV, Leiden, The Netherlands; http://www.enzyscreen.com/).
2. Deepwell microtiter plates (# CR1424, Enzyscreen BV).
3. Sandwich covers for deepwell microtiter plates (# CR1224, Enzyscreen BV).

2.2.4 Laboratory Fermenter

BIOFLO® 110 Advanced Add-A-Vessel fermenter and bioreactor kit (#s M1273-1160 and M1273-1620, New Brunswick Scientific, New Jersey, USA) equipped with the following:

1. Gas Mix controller (# M1273-3104).
2. Ex-2000 Gas Analyzer (# M1276-5000).
3. Eight Channel Sequencer (# M1154-2395).
4. Biocommand Plus computer software (# M291-0000).
5. Oil-free air compressor.

3 Methods

Carry out all procedures at room temperature unless otherwise specified. Take care for not dusting the environment with agar or other easily spreadable medium components. Autoclaving is carried out at 121 °C for 20 min if not stated otherwise. Cultivation media can also be sterilized in a pressure cooker for 30 min at 121 °C (at 15 psi).

3.1 Plate Cultures

3.1.1 Preparation of PDA Plates

1. Mix the required amount of PDA (as noted in the package label) with purified (MilliQ) water measured in a graded beaker. Fill up to 2/3 to leave room for the agar to boil in the autoclave, and cover with foil or a lid not fully closed. Sterilize at 121 °C for 20 min. There is usually no need to adjust the pH of PDA.
2. Cool the agar after autoclaving to about 60–70 °C for pouring of plates or making additions.
3. Adding Triton X-100: If required, add 10 mL of the 10 % (v/v) stock per liter of PDA after sterilization. Triton stock is sterilized separately (*see* **Note 5**). Thorough mixing can be achieved using a magnetic stirrer plate.
4. Adding Hygromycin B: Add the required amount of the antibiotic to the sterilized and well-cooled PDA so that the required final concentration in the medium will be reached. For example, add 13 μL (conc. 457,000 U/mL) per 100 mL of medium for a final concentration of 60 U Hygromycin B/mL of agar (*see* **Note 6**).
5. Cellophane discs sterilized by autoclaving as stacked between the Whatman filter paper using the dry program (e.g., 121 °C 20 min, 15-min drying) can be placed on top of the solidified agar (here PDA) using forceps flamed in 70 % (v/v) ethanol (*see* **Note 7**).

3.1.2 Preparation of Minimal Agar Plates

1. Mix 15 g KH_2PO_4 and 5 g $(NH_4)_2$ SO_4 in 700 mL of purified water in a graded beaker. Add water to agar (e.g., 20 g agar/L of water), and place the beaker on a magnetic stirrer plate to mix. Add trace elements (10 mL of 100× stock/L). Adjust the medium to the required pH (e.g., pH 5.5 or 6.5) with 5 M KOH before sterilizing and fill up to 870–920 mL with purified water. This leaves room for the addition of $MgSO_4$ and $CaCl_2$ plus the carbon source and Triton X-100 after autoclaving. Triton X-100 can be added as above.
2. Prepare a 6 % (w/v) stock from both $MgSO_4$ and $CaCl_2$ in purified water and sterilize by autoclaving. Add 100× dilution of each chemical to the sterilized agar medium (10 mL to the final volume of 1,000 mL of medium) (*see* **Note 8**).
3. Make a 10 % (w/v) stock of the required carbon source (e.g., glucose, birch wood xylan) and sterilize separately; 10 min at 121 °C or filter sterilization through a 0.22 μm filter (e.g., Steritop #SCGVT05RE, Millipore Australia Pty Ltd, Kilsyth, Victoria, Australia) is recommended for glucose to avoid caramelization. Add 10× dilution of the sugar (e.g., 100 mL to the final volume of 1,000 mL) for a final concentration of 1 % (v/v) to the agar medium sterilized separately. Dissolving a 10 % stock of a polymeric carbon source such as birch xylan may require mixing the solution on a warm magnetic stirrer plate (in this case the vial should contain a magnetic stirrer). Final concentration of the birch wood xylan in the medium is 0.5 % (w/v). For other additions, *see* **Note 2**.
4. Testing fungal colonies for xylanase production. After inoculation (*see* Subheading 3.1.4, **step 3**) and incubation of the minimal agar-birch xylan plates for the required time and temperature and Subheading 3.1.3 below, the plates will be flooded with 1.0 % (w/v) Congo red for 5–10 min. Then pour off the Congo red solution, and rinse the plates with 1 M NaCl until a yellowish clearing halo is visible around the colonies producing xylanase [6]. The halo is a result of the enzyme hydrolyzing the substrate (birch wood xylan) on the plates. The haloes should become clear in about 5–10 min (*see* **Note 9**).

3.1.3 Fungal Sporulation

1. Prepare PDA plates as in Subheading 3.1.1.
2. Thaw out fungal spore suspension kept in cryogenic storage solution at −80 °C or use fresh spore suspension prepared as in Subheading 3.1.4.
3. Spread 50–100 μL of spore suspension onto dry PDA plate (s) and spread with a sterile spreader.
4. Incubate at 28 °C for 7–10 days or until the surface of the plate is fully covered with spores.

3.1.4 Inoculation and Incubation of Plate Cultures

1. Inoculation with fungal spores: Prepare a spore suspension by pouring 5–7 mL of sterile 0.9 % NaCl (w/v) + 0.01 % Tween 80 on a PDA plate containing a premade fungal culture as in Subheading 3.1.3, and scrape the spores into solution. Filter the spore solution into a sterile 10 mL test tube through a sterile funnel containing a cotton wool plug to remove hyphae. Take 1 mL of the spore solution and dilute further (*see* **step 2** below) if required for obtaining a viable count or separate colonies, using aseptic techniques. Plate out 100 μL aliquots and spread aseptically using a sterile spreader. A glass spreader can be sterilized by flaming in 70 % (v/v) ethanol.
2. Inoculation with a soil sample: Measure 1 g of the soil sample in a test tube containing 10 mL of sterilized 0.9 % NaCl (w/v) + 0.01 % (v/v) Tween 80, and prepare a tenfold dilution series until 10^{-6}. You may need to use pipette tips of which the mouth has been widened by cutting off the end of tips. Mix well between every transfer (*see* **Note 10**).
3. Inoculation from colonies: Plates can be inoculated from colonies growing on older plate cultures. Transfer is carried out by lightly touching the growing colony by a sterile rod or toothpick and making a 1–2 mm streak onto the new plate (*see* **Note 11**). Sometimes a piece of agar containing fungal growth is cut out aseptically and placed onto a fresh plate.
4. A general incubation temperature for mesophilic fungi is +28 °C. Incubation times usually vary from 3 to 7 days depending on the fungal species. Plates will be incubated bottoms up to avoid condensation of water onto the cultures (*see* **Note 12**).

3.2 Liquid Cultures

3.2.1 Preparation of the Hydrolase-Inducing Medium

1. Weigh 1 g of Avicel cellulose and 1.5 g soybean flour (*see* **Note 13**) in a 250 mL conical shake flask.
2. Make a 10 % (w/v) stock of lactose and filter sterilize through a 0.45 μm filter.
3. Mix 15 g KH_2PO_4 and 5 g $(NH_4)_2SO_4$ with 700 mL of purified water in a graded beaker. Adjust the medium to the required pH (e.g., pH 5.5 or 6.5) with 5 M KOH before sterilizing. Optionally, a final concentration of 0.02 % (v/v) Tween 20 (# 0777, Amresco, Inc., Ohio, USA) may be added to curb foaming of medium during autoclaving and help improve permeability of the fungal cell wall. Fill up to 1,000 mL with purified water, and pour 45 mL into the flasks prepared in 1. Sterilize the flasks in an autoclave at 121 °C for 30 min.
4. Prepare a 6 % (w/v) stock for both $MgSO_4$ and $CaCl_2$ in purified water and sterilize by autoclaving. After sterilizing, add 100× dilution of each into the separately sterilized conical flasks (0.5 mL to the final volume of 50). Also add 2.5 mL of the sterilized 20 % (w/v) lactose stock in each flask. Work aseptically.

3.2.2 Inoculation and Incubation of the Standard Shake Flask Cultures

1. Prepare the spore suspension as in Subheadings 3.1.3 and 3.1.4, and count the spores under a microscope using a hemocytometer according to the instructions. Conduct inoculation of the cultures to a final concentration of your choice (e.g., 1×10^8/mL of medium). Inoculate the flasks aseptically (*see* **Note 14**).
2. Incubate the cultures at 28 °C on an orbital shaker at 250 rpm up to 7 days (*see* **Note 15**).

3.2.3 Cultivation Using the Duetz System

1. Wrap deepwell microtiter plates together with sandwich covers in foil and sterilize at 121 °C for 20 min (*see* **Note 16**).
2. Prepare sufficient cultivation medium as in Subheading 3.2.1.
3. In a laminar flow, mix the medium before aliquoting up to 4 mL per well.
4. Inoculate with 20–50 μL of spores (~1.25×10^6/mL).
5. Place the plate(s) onto a clamp cover fixed to a gyratory shaker incubator (*see* **Note 16**).
6. Incubate at 28 °C, 250 rpm, for up to 7 days.

3.2.4 Cultivation in a Laboratory Fermenter

1. Prepare a 10 % (v/v) seed culture medium as in Subheading 3.2.1 (e.g., 500 mL for a 5 L batch culture). Make the 500 mL seed in a 2 L shake flask to allow for good aeration.
2. Inoculate the medium with 3×10^4–3×10^5/mL spores and incubate at 28 °C (200 rpm) for 2–3 days. Check the seed culture for potential contamination before adding into the fermenter vessel.
3. Mix 15 g/L KH_2PO_4, 5 g/L $(NH_4)_2SO_4$, 1 mL Tween 80 (100 %), 1 mg/L $FeSO_4 \times 7H_2O$, 0.2 mg/L $MnSO_4 \times H_2O$, 0.2 mg/L $ZnSO_4 \times 7H_2O$, and 0.4 mg/L $CoSO_4 \times 7H_2O$. Add MilliRO water to 4.3 L in the fermentation vessel connected to the mixing unit. Allow for extra 50–100 mL of liquid due to evaporation during sterilization (*see* **step 9**).
4. Set agitation to 200 rpm to help mix, and dissolve salts.
5. Set pH of the medium to ~6.5 by adding in either KOH or NaOH pellets (# A482, Ajax Finechem Pty Ltd, Australia).
6. Add 100 g of Avicel cellulose and 75 g of soybean flour. Keep mixing (e.g., for 30 min) to help hydrate powders.
7. For foam control, include and sterilize 200 mL of antifoam (e.g., Struktol from Scheill & Seilacher, # SB 2023) in a separate 500 mL Schott bottle.
8. Optional: For pH control during the run, prepare 250 mL of 10 % (w/v) NaOH (base) and 250 mL of 10 % (v/v) H_3PO_4 (acid) in separate 500 mL Schott bottles.
9. All the three feed bottles (here antifoam, base, and acid) can be placed onto removable clamps normally attached to a platform at the base of the vessel (*see* **Note 17**).

10. Detach the fermentation vessel containing the medium from the mixing unit and sterilize at 121 °C for 1 h. When finished, assemble the vessel and set agitation to 250 rpm. Note: No aeration.
11. Sterilize separately 20 % (w/v) lactose (250 mL), 1 M $MgSO_4$ (12 mL), and 1 M $CaCl_2$ (27 mL), and add each component into the vessel aseptically. The total working volume is ~5 L with a seed of 500 mL.
12. Leave the vessel for up to 2 days at RT, and check for contamination under the microscope before addition of seed from **step 2**.
13. When ready and before addition of seed, first set the vessel temperature to 28 °C and equilibrate dissolved oxygen (DO) to ~100 % by increasing agitation to 500 rpm and aeration to 10 standard liter per minute (SLPM).
14. Once the DO has been equilibrated to ~100 %, change agitation down to 250 rpm (TPSD = 96.81 cm/s) and aeration to 2 SLPM (0.14 VVM). If required, manually add in antifoam using one of the ports provided in the vessel top working aseptically to remove foam buildup during equilibration. Set the fermentation program for DO to a minimum of 10 % with "cascade" linked to an agitation span of 250–500 rpm.
15. Inoculate the fermenter vessel with the seed using an appropriate inoculation port, working aseptically. Start the computer software (e.g., Biocommand Plus) to track and keep records of the different running parameters (e.g., DO, pH, agitation, and exhaust gases CO_2 and O_2).
16. Remove a sample through the sampling port (e.g., 50 mL = log0) and keep at 4 °C until analyzed.
17. Remove samples daily (up to day 7), check for contamination, and keep at 4 °C for further analysis.
18. Once the run is completed, prepare the vessel for the final harvesting of the culture, for example, by centrifugation. Clarify the samples taken each day by centrifugation (e.g., 17,000 × *g* for 30 min), and use clear supernatants for analysis (SDS-PAGE, 2D electrophoresis, enzyme activity, etc.) of secreted gene products of interest.

4 Notes

1. It is important to choose non-coated cellophane since coated brands do not allow adsorption of nutrients through the discs, thus preventing fungal growth. Suitable cellophane can be purchased from a news agency, for example.
2. Examples of carbohydrate polymer degrading enzymes secreted by saprophytic soil fungi include cellulases and starch-degrading enzymes such as amylases and glucoamylases.

Table 1
Substrates that can be introduced into growth plates to indicate production of a particular hydrolase activity

Enzyme activity	Substrate added	References
Amylase	Soluble starch	[7]
Cellulase	Walseth cellulose	[8, 9]
	Carboxymethyl cellulose	[10]
	Hydroxyethyl cellulose	[11]
Chitinase	Chitin	[7, 12]
DNAse	Deoxyribonucleic acid	[7]
β-Glucanase	AZCL-pachyman	[13]
β-Glucosidase	Esculin and FAC	[10]
Lipase	Sodium monolaurate (Tween 20)	[7]
	Olive oil and rhodamine B	[14]
β-Mannanase	Locust bean gum	[15]
	OBR-galactomannan	[16]
Pectinase	Pectin (citrus or apple)	[7, 11]
Phosphatase	Phenolphthalein diphosphate (sodium salt)	[7]
Protease	Gelatin	[7, 11]
	Skim milk	[17]
RNAse	Ribonucleic acid (yeast)	[7]
Urease	Urea	[7]
Xylanase	Birch xylan	[18]
	RRB-xylan	[16]

Esculin 7,6-dihydro-oxycoumarin-6-glucoside, *FAC* ferric ammonium citrate, *OBR-galactomannan* ostazin brilliant red-galactoglucomannan, *RRB-xylan* remazol brilliant blue-xylan

These enzyme activities, produced by growing hyphae, can be visualized on cultivation plates where, typically, the carbon source provided is a substrate for a particular enzyme. On plate screening, enzyme production is often indicated by formation of a large clearing zone (halo) around the growing colony or by emergence of a colored or a fluorescing product (depending on the substrate used). Here we provide an example of using birch wood xylan for the screening of xylanase activity assessed by formation of a hydrolysis halo around the colonies growing on the plate containing birch wood xylan as the carbon source. *See* Table 1 above for other suggestions for substrates to be added to the growth media for the screening of different enzyme activities.

3. Petri dishes used for fungal cultures are typically higher than those used for bacteria. This is to accommodate the fungal colonies that form aerial structures such as conidiophores. The plates are also vented to provide proper aeration of the cultures still keeping them sterile.
4. It is recommended to use plastic or metal caps instead of cotton bungs that may collect moisture during cultivation. Shake cultures for initial testing of, e.g., enzyme activities can be carried out in test tubes keeping in mind that the aeration is not the same as in 250 mL conical flasks containing 50 mL medium.
5. Sterile Triton X-100 (10 % stock/L of agar) can be added to any agar medium including minimal agar plates (*see* below). Triton X-100 is used to restrict the growth of fungal colonies on the plates. This is especially handy when several colonies must be accommodated on one plate.
6. Any antibiotic can be added to an autoclaved and cooled agar medium at the required concentration. It is not advisable to autoclave antibiotic solutions which are usually prepared by adding sterile water to a powder or using an antibiotic purchased in a ready-made solution. Note that the antibiotic can also be added to the overlay agar when selecting for, e.g., Hygromycin B-resistant transformant colonies that are first grown on PDA without an antibiotic from 4 h to overnight.
7. Fungal hyphae are often cultivated on top of cellophane discs in cases where direct contact of the growing hyphae with the agar medium is not desired. Such situations include growing hyphae for the isolation of DNA [19] or RNA (e.g., using the Trizol® reagent; Invitrogen, Australia) and some biocontrol experiments where the interacting strains need to be kept separate. Typically, DNA or RNA will be isolated from lyophilized mycelia scraped off from the cellophane disc and ground into fine powder under liquid nitrogen.
8. Addition of $MgSO_4$ and $CaCl_2$ before autoclaving would cause precipitation.
9. Flooding with 1 M NaCl will sharpen the hydrolysis halos, but continuing for too long will change the pH of the agar and cause it to turn blue-black. Size of the halo is indicative of the amount of enzyme produced/secreted by the fungal colony. Since the flooding chemicals are not sterile, a master plate (e.g., PDA) with colonies corresponding to those on the screening plate is required.
10. Making a dilution series can be applied when plating out samples taken from other types of sources such as other organic material or water. High dilutions usually aid in obtaining separate colonies starting from individual spores. This helps in isolation of the type of fungus of interest and is also a good method for purifying fungal cultures through single spores.

For metagenomic studies, DNA can be extracted directly from soil samples using the method of [20] without any prior plating of the material.

11. When streaking colonies on the screening plates it is important to keep the streaks short enough to avoid the developing colonies growing together. This would complicate interpretation of the clearing halos, etc.
12. The temperature range for fungal growth is +20–42 °C. It is worth a note that some fungal colonies produce a large amount of spores that may spread all over the plate if the plate is tapped against a bench, for example. This is especially damaging if the goal is to produce culture containing one species only.
13. Some batches of soybean flour may contain bacterial spores as a contamination. If this is suspected, it is advisable to autoclave the medium (e.g., 50 mL culture in 250 mL conical shake flask) for a prolonged time, e.g., 30 min at 123 °C (or sterilize the medium twice for 30 min at 121 °C). Also, checking a sample of the culture under the microscope for potential contamination during the cultivation is recommended.

    We have described here only one type of cultivation medium; however, the Avicel cellulose can be replaced by any other type of cellulose or other carbon source and the soybean flour with any other desired (complex) nitrogen source. Liquid cultivations can also be carried out using minimal medium prepared according to the principles described in Subheading 3.1.2 by just leaving out the solidifying agent.
14. The inoculum size may have an effect on the productivity of the culture. The correct size is usually found by trial and error and can be applied to both shake flask cultures (Subheading 3.2.1) and the Duetz system (Subheading 3.2.3), whereby the culture starts directly from spores and not relying on a 2–3-day seed as in laboratory-scale fermentation (Subheading 3.2.4). The final concentration of the spores to be inoculated in a liquid culture can be in the range of 10^5–10^8/mL. It should be noted that as the mode of growth of fungi is filamentous, measurement of the OD to assess growth density is not reliable even though it has been used on a small scale [21].
15. It is recommended to incubate the fungal cultures in the dark as daylight may induce sporulation of certain fungi. Superfluous sporulation may also affect a negative effect of the enzyme yields.
16. Assembly of the inoculated Duetz deepwell microtiter plates ready to be placed, e.g., on a shaker for cultivation (Fig. 1a). The deepwell microtiter plates (opaque white) together with sandwich covers equipped with holes for aeration (Fig. 1b).
17. Bench top fermenter (Fig. 2). Feed bottles can be placed on removable clamps attached to the base of the fermenter vessel. Inoculation ports are located at the top plate.

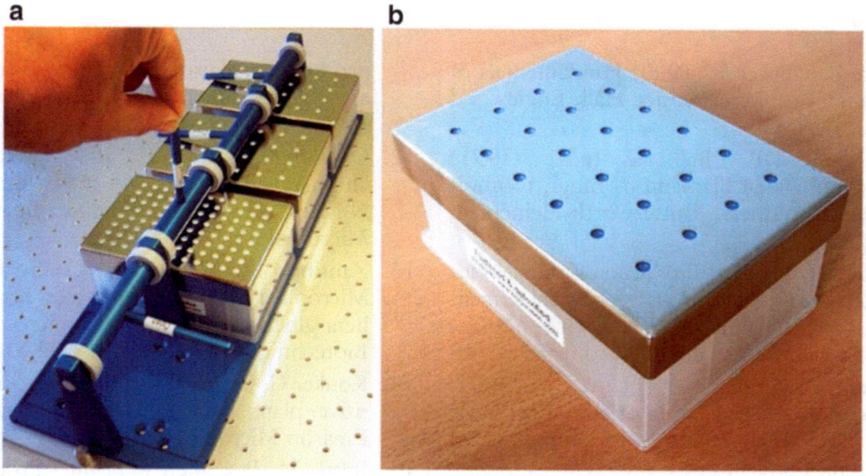

Fig. 1 Duetz system for cultivation of fungi. (**a**) Platform containing different types of microtiter plates (MTP) for cultivation of microorganisms including fungi. (**b**) A closeup and magnified view of a 24-deepwell MTP with lid. With permission from Enzyscreen BV (http://www.enzyscreen.com/)

Fig. 2 Fermentation setup for controlled and submerged cultivation of microorganisms such as filamentous fungi, with a capacity of up to 10 L working volume. The picture demonstrates the fermenter control panels on the *left* with the glass vessel for cultivation on the *right*

References

1. Moore-Landecker E (1990) Fundamentals of the fungi, 3rd edn. Prentice Hall, Englewood Cliffs, NJ, 561
2. Finkelstein DB, Ball C (eds) (1991) Biotechnology of filamentous fungi, technology and products. Butterworth-Heineman, MA, p 520
3. Brand D et al (2010) Production of fungal biological control agents through solid state fermentation: a case study of *Paecilomyces lilacinus* against root knot nematodes. Mico Int 22:31–48
4. Hölker U, Höfer M, Lenz J (2004) Biotechnological advances of laboratory-scale solid-state fermentation with fungi. Appl Microbiol Biotechnol 64:175–186
5. Duetz WA (2007) Microtiter plates as mini-bioreactors: miniaturization of fermentation methods. Trends Microbiol 15:470–475
6. Teather RM, Wood PJ (1982) Use of Congo red polysaccharide interactions in enumeration and characterization of cellulolytic bacteria from the bovine rumen. Appl Environ Microbiol 43:777–780
7. Hankin L, Anagnostakis SL (1975) The use of solid media for detection of enzyme production by fungi. Mycologia 47:597–607
8. Walseth CS (1952) Influence of fine structure of the cellulose on the action of cellulases. TAPPI 35:233–238
9. Bailey MJ, Nevalainen KMH (1981) Induction, isolation and testing of stable *Trichoderma reesei* mutants with improved production of solubilizing cellulase. Enzyme Microb Technol 3:153–157
10. Melo IS, Faull JL, Graeme-Cook KA (1997) Relationship between in vitro cellulose production of UV-induced mutants of *Trichoderma harzianum* and their bean rhizosphere competence. Mycol Res 101:1389–1392
11. Hagerman AE, Blau DM, McClure AL (1985) Plate assay for determining the time of production of protease, cellulose and pectinase by germinating fungal spores. Anal Biochem 151:334–342
12. O'Brien M, Colwell RR (1987) A rapid test for chitinase activity that uses 4-methylumbelliferyl-N-acetyl-β-D-glucosaminidine. Appl Environ Microbiol 53:1718–1720
13. Zantinge JL, Huang HC, Cheng KJ (2002) Microplate diffusion assay for screening of beta-glucanase-producing micro-organisms. Biotechniques 33:798, 800, 802 passim
14. Kouker G, Jaeger K-E (1987) Specific and sensitive plate assay for bacterial lipases. Appl Environ Microbiol 53:211–213
15. Rättö M, Poutanen K (1988) Production of mannanase-degrading enzymes. Biotechnol Lett 10:661–664
16. Kremnický L et al (1996) Production of extracellular β-mannanases by yeasts and yeast-like microorganisms. Folia Microbiol 41:43–47
17. Mäntylä A et al (1994) Cloning of the aspartic protease gene of *Trichoderma reesei*. In: The second European conference on fungal genetics, Lunteren, The Netherlands, Abstract B52
18. Kubata BK et al (1994) Purification and characterization of *Aeromonas caviae* ME-1 xylanase V, which produces exclusively xylobiose from xylan. Appl Environ Microbiol 60:531–535
19. Lee SB, Milgroom MG, Taylor JW (1988) A rapid, high yield mini prep method for isolation of total genomic DNA from fungi. Fungal Genet Newsl 35:23–24
20. Vázquez-Marrufo G et al (2002) DNA isolation from forest soil suitable for PCR assays of fungal and plant rRNA genes. Plant Mol Biol Rep 20:379–390
21. Chin-A-Woeng TFC et al (2002) Visualisation of rhizosphere interactions of *Pseudomonas* and *Bacillus* biocontrol strains. In: Varma A, Abbott L, Werner D, Hampp R (eds) Springer-Verlag, Berlin, pp 431–448

Chapter 2

Rapid Extraction of PCR-Competent DNA from Recalcitrant Environmental Samples

Michael R. Gillings

Abstract

Advances in sequencing technologies have made the investigation of microbial ecology and community dynamics more tractable. The critical first step in such analyses is the efficient and representative recovery of PCR-competent DNA from complex environmental samples. All extraction protocols contain inherent biases, meaning that choice of method involves compromise between various factors, including efficiency, yield, universality, and representative extraction. Here, details are given for a routine method used in our laboratory to extract DNA from soils, sediments, biofilms, roots, and fungi.

Key words Microbial diversity, PCR, Microbial ecology, Soil, Sediment, Biofilm

1 Introduction

Our understanding of microbial ecology and diversity has rapidly expanded since the introduction of molecular methods for studying microbial communities [1], and the pace at which our knowledge accumulates will accelerate with the widespread use of next-generation sequencing technologies [2]. Because the majority of microorganisms are yet to be cultured [1], investigations must rely on direct analysis of environmental samples. A typical first step is the direct extraction of total DNA from environmental samples, which often contain diverse phyla and interfering substances in a complex matrix. The problems of representative extraction of DNA from complex substrates such as soil have been the subject of many studies.

Effective methods for DNA extraction must generate good yields of DNA, they must be unbiased in terms of recovering DNA from the diverse species that are present, and they must generate DNA that is suitable for downstream applications [3–5]. Many different methods for extraction of DNA from environmental samples have been published, but it appears that there is no one method that is free from bias [6–10]. Consequently, researchers can opt for

Ian T. Paulsen and Andrew J. Holmes (eds.), *Environmental Microbiology: Methods and Protocols*, Methods in Molecular Biology, vol. 1096, DOI 10.1007/978-1-62703-712-9_2, © Springer Science+Business Media, LLC 2014

a specialized method tailored to their particular environment or can make a pragmatic choice based on a method's wide applicability to a number of systems.

Of available methods, those involving physical disruption by bead beating appear to have the broadest applicability and are readily available in kit form. Since the first description of this approach [11], various studies have shown that bead beating generates good yields of DNA [12] and that it retrieves DNA from a wide diversity of organisms, such that it is useful for comparative studies [13–15]. Bead beating can be optimized for particular applications [16] and is the basis of the lysis procedure used in the international standard soil extraction method ISO 11063 [17].

Here an outline is given for one rapid and adaptable DNA extraction method. It is based on a commercial kit (FastDNA, MP Biomedicals) that employs lysis with bead beating and DNA purification by absorption to silica. The protocol was originally described by Borneman et al. [11] and has been modified to speed up the extraction process and reduce reliance on proprietary reagents [18]. This method is rapid, generates DNA suitable for PCR analysis, and avoids the use of hazardous reagents. It is also adaptable and is able to extract DNA from a wide range of environmental samples including soils with different clay contents and soils polluted with heavy metals and aromatics [18]. The method has been used to extract DNA from soils for analysis of fungal diversity [19] and to extract DNA from fungal cultures [20], lichens, and mushrooms. It also works effectively on marine and freshwater biofilms [21–23] and on diverse sediments, charcoal filters and fecal samples [24, 25]. With minor modifications it can be used to extract DNA from plant roots and their associated microbiota and from bacterial spores and gram-positive organisms that are otherwise difficult to lyse [26].

2 Materials

Solutions are made with distilled water that has been 0.22 μM filtered and then autoclaved. All user-prepared solutions are autoclaved again prior to addition of ethanol or SDS as required. All plasticware is also sterilized by autoclaving.

1. Environmental sample (*see* **Note 1**).
2. Lysing matrix E tubes (MP Biomedicals).
3. Balance.
4. FastPrep bead-beating machine (MP Biomedicals).
5. Sodium phosphate buffer (100 mM, pH 7.0).
6. MT buffer (MP Biomedicals proprietary reagent) (1 % sodium dodecyl sulfate, 1 % polyvinylpyrrolidone 40, EDTA plus proprietary inorganic salts) (*see* **Note 2**).

7. CLS-VF buffer (MP Biomedicals proprietary reagent) (sodium dodecyl sulfate, polyvinylpyrrolidone 40, Teepol 610S, and EDTA) (*see* **Note 2**).
8. CLS-TC buffer (MP Biomedicals proprietary reagent) (urea, sodium phosphate, SDS, and dithiothreitol) (*see* **Note 2**).
9. Microcentrifuge.
10. Protein precipitation solution (3 M potassium acetate, 4 % glacial acetic acid).
11. Vortex machine.
12. Binding matrix (MP Biomedicals glassmilk). Can be diluted 1:5 (v/v) with 6 M guanidine isothiocyanate. Store at 25 °C or above (*see* **Note 3**).
13. Wash buffer (100 mM sodium acetate, 70 % v/v ethanol).
14. Rotator wheel (optional).
15. TE buffer (10 mM Tris–HCl pH 8.0, 1 mM EDTA) (*see* **Note 4**).

3 Methods

3.1 DNA Extraction

Samples are homogenized by bead beating, proteins and polysaccharides selectively precipitated, and DNA purified by absorption onto glassmilk.

3.1.1 DNA from Soils, Sediments, Biofilms, and Fungi

1. Weigh 200–400 mg of environmental sample (soil or sediment) into a lysing matrix E tube. For biofilms, fecal material, lichens, mushrooms, or fungal hyphae, a sample of 100–200 mg should be used. In the case of diffuse biofilms or planktonic cells, these can be collected by centrifugation and resuspended in 780 μl of phosphate buffer before transfer into the lysing matrix tube.
2. Add 780 μl of phosphate buffer and 122 μl of MT buffer.
3. Tighten the screw-cap on the tube, making sure that no sample is trapped between the rim and the internal O-ring.
4. Load tubes into the FastPrep machine and process for 30 s at 5.5 m/s. If some material still looks unhomogenized, process for a further 30 s after waiting 1 min for the tubes to cool down. If the original samples were dry, leave the tubes to incubate at room temperature for 15 min to 2 h before proceeding to centrifugation. This improves extraction efficiency.
5. Centrifuge for 5 min at 14,000 × *g* to pellet beads and soil debris.
6. Recover 600 μl of supernatant into a fresh 1.5 ml tube (*see* **Note 5**).
7. Add 150 μl protein precipitation solution and gently vortex to mix. Stand at room temperature for a few minutes to allow the precipitate to form.

8. Centrifuge for 5 min at 14,000 × *g* to pellet proteins and polysaccharides.
9. Recover 700 μl of the supernatant into a fresh 1.5 ml tube (*see* **Note 5**).
10. Add 700 μl of binding matrix and mix thoroughly with the supernatant. Tubes may be placed on a rotator for 5 min to increase the efficiency of DNA binding.
11. Pulse spin the tubes to pellet the glassmilk, now containing the bound DNA.
12. Decant the supernatant, and gently resuspend the pelleted glassmilk in 800 μl of wash buffer. Tubes may be placed on a rotator for 5 min to increase the efficiency of washing.
13. Pulse spin the tubes to pellet the washed glassmilk, and decant the used wash buffer. If the wash buffer is brown or colored, **step 12** may be repeated.
14. Place the tubes back in the centrifuge and pulse spin to collect any remaining supernatant in the bottom of the tube.
15. Carefully remove the remaining supernatant with a micropipette.
16. Air-dry the pellet for a few minutes.
17. Resuspend the glassmilk pellet in 200 μl of TE buffer. Ensure that the pellet is fully resuspended, and allow the TE to elute the DNA at room temperature for 5 min. Elution can be improved by incubation at 50 °C in a water bath or heat block.
18. Centrifuge for 2 min at 14,000 × *g*.
19. Recover 160 μl of the supernatant, which now contains the environmental DNA.
20. Transfer to a fresh, labeled tube, and store at −20 °C until use.

This method works well for soil samples, marine and freshwater sediment samples, biofilms, and feces. It can also be used for extracting DNA from pure fungal cultures grown over cellophane on agar plates or from mushrooms, toadstools, and lichens.

3.1.2 DNA from Plant Roots and Leaves

For extracting DNA from leaves or plant roots (including plant symbionts) the following steps can be substituted for **steps 1–8** above:

1. Finely chop 200 mg of plant material with a sterile scalpel and add to a lysing matrix E tube.
2. Add 800 μl of CLS-VF buffer and 200 μl of protein precipitation solution.
3. Load tubes into the FastPrep machine and process for 30 s at 5.5 m/s. If some material still looks unhomogenized, process for a further 30 s after waiting 1 min for the tubes to cool down.
4. Centrifuge at 14,000 × *g* for 5 min to pellet beads, cell debris, polysaccharides, and proteins.
5. Continue from **step 9** above.

3.1.3 DNA from Gram-Positive Organisms and Spores

For cultures of gram-positive organisms and spore formers, the following modification can be used.

1. Resuspend 40 mg of bacterial culture in 1,000 μl of CLS-TC buffer and transfer to a lysing matrix tube. If crystals have formed in the CLS-TC, warm the solution in a water bath at approximately 50 °C to redissolve the urea.
2. Load tubes into the FastPrep machine and process for 40 s at 6.0 m/s.
3. Centrifuge at 14,000 × *g* for 5 min to pellet beads, cell debris, polysaccharides, and proteins.
4. Continue from **step 9** above.

3.2 Quality Control

The yield and purity of DNA can be qualitatively tested by agarose electrophoresis and by performing PCR directed at universal target genes.

1. Load an aliquot (10 μl) of the DNA prepared in Subheading 3.1 onto a 1 % w/v agarose gel, and subject it to electrophoresis. Stain and photograph the gel according to standard procedures.
2. Examine the molecular weight of the extracted DNA in comparison with a molecular weight marker or DNA ladder. Extracted DNA should be larger than 10 kb and should be visible as a coherent band without smearing into low-molecular-weight regions. No RNA should be present in the extraction, but if it is, RNase A digestion can be performed concurrently with the PCR by adding 1 μl of a 1 mg/ml pre-boiled RNAse A solution to each reaction.
3. Estimate the amount of DNA present by comparison to known amounts present in the molecular weight ladder. It is often not possible to obtain accurate spectrophotometer readings from environmental DNA because of interfering substances.
4. Use 1 μl of the DNA and 1 μl of a 1:10 dilution in TE as templates for PCR. The test PCR should be directed at a high-copy-number DNA target that is known to be present in the original environmental sample. The 16S ribosomal RNA genes are a suitable target for bacterial DNA, and any set of universal 16S rDNA primers can be used in this test [18]. For eukaryotic targets, the internal transcribed spacer region (ITS) of the ribosomal RNA genes is also a good target for universal primers.
5. Run out the PCR on a 2 % (w/v) agarose gel according to standard methods. Most extractions will generate amplicons from both neat and diluted DNA; however, if PCR inhibitors are present, amplicons may only be generated from the 1:10 dilution. If no amplification is generated despite the presence of visible DNA, then more washing of the binding matrix (**step 12**) may be warranted. Otherwise further dilution or

cleanup of the DNA using ethanol precipitation may be needed. It is unusual for DNAs prepared using the technique in Subheading 3.1 to not be PCR competent.

4 Notes

1. Sampling strategies for soils, sediments, and biofilms should be carefully planned and tailored towards the hypothesis being tested. Consideration should be given to both small- and large-scale spatial variation as well as temporal variation and appropriate replicates taken. Ideally, samples should be processed with minimal delay to maintain the microbial assemblages present in the field. Storage of temperate samples at 4 °C is recommended to preserve communities prior to extraction.
2. While most reagents can be prepared by the user, the exact formulation of MT, CLS-VF, and buffers are not known, and these are best purchased from the supplier.
3. Because DNA yields from environmental samples are usually low, they fall far below the binding capacity of the glassmilk, which can be diluted without appreciable loss of yield. Warm the solution to redissolve any crystals that might form, and resuspend the glassmilk immediately before use.
4. We store TE buffer and sterile PCR water as 1 ml aliquots frozen at −20 °C. After thawing for use, any unused portions are disposed of. This procedure helps to manage potential contamination.
5. The volumes of supernatant that are harvested can be increased, but the relative proportions of the solutions added in the next step must be also proportionally increased.

References

1. Hugenholtz P, Goebel BM, Pace NR (1998) Impact of culture-independent studies on the emerging phylogenetic view of bacterial diversity. J Bacteriol 180:4765–4774
2. Sogin ML et al (2006) Microbial diversity in the deep sea and the underexplored "rare biosphere". Proc Natl Acad Sci USA 103:12115–12120
3. Frostegard A et al (1999) Quantification of bias related to the extraction of DNA directly from soils. Appl Environ Microbiol 65: 5409–5420
4. Krsek M, Wellington EMH (1999) Comparison of different methods for the isolation and purification of total community DNA from soil. J Microbiol Methods 39:1–16
5. Roose-Amsaleg CL, Garnier-Sillam E, Harry M (2001) Extraction and purification of microbial DNA from soil and sediment samples. Appl Soil Ecol 18:47–60
6. Feinstein LM, Sul WJ, Blackwood CB (2009) Assessment of bias associated with incomplete extraction of microbial DNA from soil. Appl Environ Microbiol 75:5428–5433
7. Inceoglu O et al (2010) Effect of DNA extraction method on the apparent microbial diversity of soil. Appl Environ Microbiol 76: 3378–3382
8. Martin-Laurent F et al (2001) DNA extraction from soils: old bias for new microbial diversity analysis methods. Appl Environ Microbiol 67:2354–2359

9. Robe P et al (2003) Extraction of DNA from soil. Eur J Soil Biol 39:183–190
10. Thakuria D et al (2008) Importance of DNA quality in comparative soil microbial community structure analyses. Soil Biol Biochem 40:1390–1403
11. Borneman J et al (1996) Molecular microbial diversity of an agricultural soil in Wisconsin. Appl Environ Microbiol 62:1935–1943
12. Lakay FM, Botha A, Prior BA (2007) Comparative analysis of environmental DNA extraction and purification methods from different humic acid-rich soils. J Appl Microbiol 102:265–273
13. Carrigg C et al (2007) DNA extraction method affects microbial community profiles from soils and sediment. Appl Microbiol Biotechnol 77:955–964
14. de Lipthay JR et al (2004) Impact of DNA extraction method on bacterial community composition measured by denaturing gradient gel electrophoresis. Soil Biol Biochem 36:1607–1614
15. Luna GM, Dell'Anno A, Danovaro R (2006) DNA extraction procedure: a critical issue for bacterial diversity assessment in marine sediments. Environ Microbiol 8:308–320
16. Bürgmann H et al (2001) A strategy for optimizing quality and quantity of DNA extracted from soil. J Microbiol Methods 45:7–20
17. Petric I et al (2011) Inter-laboratory evaluation of the ISO standard 11063 "Soil quality—method to directly extract DNA from soil samples". J Microbiol Methods 84:454–460
18. Yeates C, Gillings MR (1998) Rapid purification of DNA from soil for molecular biodiversity analysis. Lett Appl Microbiol 27:49–53
19. Green JL et al (2004) Spatial scaling of microbial eukaryote diversity. Nature 432:747–750
20. Stow A et al (2010) Differential antimicrobial activity in response to the entomopathogenic fungus Cordyceps in six Australian bee species. Aust J Entomol 49:145–149
21. Gillings MR, Holley MP, Selleck M (2006) Molecular identification of species comprising an unusual biofilm from a groundwater treatment plant. Biofilms 3:19–24
22. Holmes AJ et al (2001) Phylogenetic structure of unusual aquatic microbial formations in Nullarbor caves, Australia. Environ Microbiol 3:256–264
23. Wilson GS et al (2010) Heterogeneity of surface attached microbial communities from Sydney Harbour, Australia. Marine Genomics 3:99–105
24. Gillings M et al (2008) The evolution of class 1 integrons and the rise of antibiotic resistance. J Bacteriol 190:5095–5100
25. Gillings MR et al (2008) Recovery of diverse genes for class 1 integron-integrases from environmental DNA samples. FEMS Microbiol Lett 287:56–62
26. Young A et al (2006) Genetic uniformity among international isolates of Leifsonia xyli subsp. xyli, causal agent of ratoon stunting disease of sugarcane (Saccharum interspecific hybrids). Aust Plant Pathol 35:503–511

Chapter 3

Quantitative PCR for Detection of mRNA and gDNA in Environmental Isolates

Anthony J. Brzoska and Karl A. Hassan

Abstract

Quantitative PCR is used to gauge the abundance of specific nucleic acid species within purified samples. Due to its high sensitivity and minimal operation costs, this method is routinely applied in modern molecular bioscience laboratories. Nonetheless, all quantitative PCR experiments must include several carefully designed, yet simple, controls to ensure the reliability of the analyses. The aim of this chapter is to provide basic quantitative PCR methods, from primer design through data analysis, that are generally applicable to studies in microbiology. These methods allow the abundance of targeted RNA or DNA molecules to be determined in nucleic acid samples purified from a variety of biological sources.

Key words Quantitative, PCR, Reverse transcription, Delta-delta C_T, Amplification efficiency

1 Introduction

Quantitative PCR (qPCR) is a versatile technique employed to determine the abundance of specific DNA species within a sample. Coupled with reverse transcription (RT) in qRT-PCR this method can also be used to gauge the relative abundance of RNA molecules. These methods are based on the principle that the number of copies of a targeted DNA sequence doubles in a typical PCR until one or more reagents become limiting. Therefore, the cycle at which a detectable product appears (the cycle threshold; C_T) is related to the starting amount of template (Fig. 1). Due to the limited expense of qPCR experiments, they are within reach of most laboratories and have become routine. However, several factors, including template purity and amplification efficiency, can have a profound effect on the results obtained, and the establishment of careful controls is mandatory to ensure that the data adhere to the assumptions of the analysis [1]. Despite the importance of these control reactions, the application of qPCR methods need not be daunting, and successful experiments are easily within the capabilities of anyone with basic molecular biology training.

Ian T. Paulsen and Andrew J. Holmes (eds.), *Environmental Microbiology: Methods and Protocols*, Methods in Molecular Biology, vol. 1096, DOI 10.1007/978-1-62703-712-9_3, © Springer Science+Business Media, LLC 2014

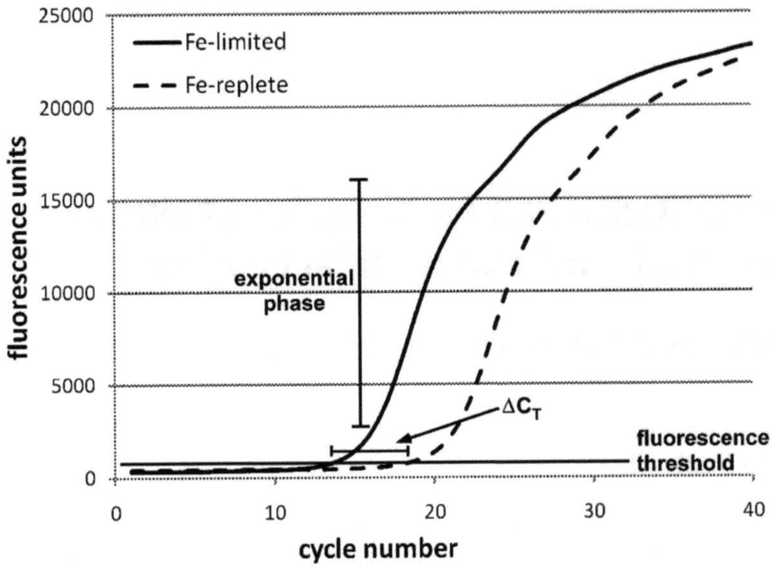

Fig. 1 Amplification of the *Acinetobacter baumannii* putative siderophore export gene, *aedD*, from cDNA of cells grown under iron-limited (*dashed line*) and iron-replete (*solid line*) conditions [13]. C_T values are determined as the point at which the fluorescence level of a reaction increases above the background level and crosses the fluorescence threshold. Amplification curves for the internal control gene, *GAPDH*, under iron-limited and iron-replete conditions are not shown, but were equivalent to each other and approximately follow the *aedD* iron-replete curve

Since its inception, qPCR has been developed into a highly sophisticated technique, offering a range of methodological options that can influence the success of the detection experiment [2]. Current qPCR technologies employ fluorescence reporters that enable PCR products to be detected in real-time on thermal cyclers equipped with fluorescence optics. These technologies are very sensitive and allow monitoring of target molecule accumulation throughout the exponential amplification phase (Fig. 1). Several chemistries are available for fluorescence detection. The most commonly used chemistries, for which there are several alternative suppliers (Subheading 2), include dyes that fluoresce upon intercalation with double stranded DNA (dsDNA; e.g., SYBR Green) [3], and sequence-specific reporter probes (e.g., TaqMan probes and molecular beacons) [4, 5].

Here we outline a complete qPCR protocol for the detection of RNA species within a cultured sample using a dsDNA-intercalating fluorescence reporter dye. The qPCR primer design protocol (Subheading 3.1) is applicable to the detection of either DNA or RNA species. Subheadings 3.2 and 3.3 describe the purification and reverse transcription of RNA to cDNA and should be followed for RNA transcript detection. For the detection of DNA markers in cultured or environmental samples, the DNA extraction procedures outlined in Subheading 3.4 should be followed.

The qPCR protocol described in Subheading 3.5 and the efficiency controls presented in Subheading 3.6 are common to both DNA and RNA detection. Finally, Subheadings 3.7 and 3.8 describe alternative methods for the calculation of DNA or RNA abundance in nucleic acid samples.

2 Materials

2.1 General Materials

1. Samples of starting materials containing target organisms or DNA template prepared from target organisms or cDNA template synthesized from mRNA extracted from target organisms.
2. Growth media.
3. RNAse-free, molecular biology grade water.
4. Forward and reverse oligonucleotide primer mix specific to target sequences, 1.3 μM each (available from local oligonucleotide supply companies).
5. Microcentrifuge and tabletop centrifuge with refrigeration unit.
6. Vortex mixer.
7. Nuclease-free plasticware; 1.5 mL microcentrifuge tubes; 50 mL conical tubes; 0.2 μL or 0.5 μL thin-walled PCR tubes; 0.2 μL PCR strip tubes or 96 well PCR microplates (for larger-scale qPCR applications).
8. Real-time thermal cycler (suppliers include: Eppendorf International, Hamburg, Germany; Applied Biosystems Incorporated, USA; Roche Applied Science, Switzerland).
9. NanoDrop™ Spectrophotometer, for the determination of DNA and RNA concentration (Thermo Fisher Scientific Incorporated, USA).
10. Spectrophotometer to monitor cell growth.
11. 2× SYBR® Green (or equivalent fluorescent DNA intercalating dye) Master Mix (examples include: GoTaq® qPCR Master Mix, Promega Corporation, USA; SYBR® Advantage qPCR Premix, Clontech Laboratories Incorporated, USA; Platinum® SYBR® Green qPCR SuperMix-UDG, Invitrogen Life Technologies Corporation, USA).

2.2 Materials for the Purification of Total RNA

1. Commercial kit for the extraction of total RNA (examples include: PureLink™ Micro-to-Midi™ Total RNA Purification System, Invitrogen Life Technologies Corporation, USA; SV Total RNA Isolation System, Promega Corporation, USA; Aurum total RNA mini kit, Bio-Rad Laboratories, USA).
2. TRIzol® Reagent (Invitrogen Life Technologies Corporation, USA). *Optional*: RNA*later* (Ambion Incorporated, Applied Biosystems, Life Technologies, USA), for the preservation of RNA species; to be used when the determination of mRNA transcript half-lives are an experimental consideration.

3. Chloroform (Molecular Biology Grade).
4. RNAse-free DNase enzyme for the preparation of DNA-free RNA (examples include: TURBO™ DNase, Ambion Incorporated, Applied Biosystems, Life Technologies, USA; Dnase I (RNAse-free), New England Biolabs, USA; Recombinant DNase I, Roche Applied Science, Switzerland).

2.3 Materials for the Synthesis of cDNA

1. Reverse transcriptase enzyme for the synthesis of cDNA from isolated RNA molecules (examples include: SuperScript® VILO™ cDNA Synthesis Kit, Invitrogen Life Technologies Corporation, USA; iScript™ cDNA Synthesis Kit, Bio-Rad Laboratories, USA; QuantiTect® Reverse Transcription Kit, Qiagen Bioscience Corporation, USA).

2.4 Materials for the Purification of Genomic DNA

1. Commercial kit for the extraction of genomic DNA from environmental organisms (examples include: ISOLATE Genomic DNA mini-kit, Bioline Reagents Limited, USA; illustra™ Bacteria genomicPrep Mini Spin Kit, GE Healthcare LifeSciences, USA; QIAamp DNA Mini Kit, QIAGEN Pty. Ltd., Hilden, Germany; MagPrep® Bacterial Genomic DNA Kit, Novagen, Darmstadt, Germany).

3 Methods

3.1 Primer Design

Primer3 is a Web-based application for the design of PCR primers, available to users free of charge [6]. This section describes the design of primers featuring appropriate characteristics for successful qPCR using Primer3-based software. For primer design involving more than a few sequences, the high-throughput Primer3-derived program BatchPrimer3 may be used [7]. BatchPrimer3 can aid in the design of primers for up to 500 candidate sequences submitted in a single input file. A representative study using high-throughput BatchPrimer3 primer design can be found in reference [8].

1. Select target genes or sequences. For the assessment of relative mRNA transcript expression, choose an appropriate endogenous control gene to use for the calculation of differential gene expression in experimental samples (*see* **Note 1**). Quantification of copies of a gene target within an environmental sample or a mixed population of bacterial cells does not require an endogenous control gene for the normalization of qPCR data sets (Subheading 3.8).
2. Download sequence data from NCBI (http://www.ncbi.nlm.nih.gov), KEGG (http://www.genome.jp/kegg/), or other appropriate genome database, for the sequence or group of sequences under study.

   For primer design to individual sequences, *see* **step 3(a)**. For high-throughput primer design, *see* **step 3(b)**:

Individual sequences:

3. (a) Open the Primer3 Web page (http://frodo.wi.mit.edu/primer3/) and paste sequence data into the source sequence field. Once completed, continue the primer design process from **step 4** onwards.

High-throughput primer design:

(b) Create a text file (.txt extension) of downloaded study sequences in FASTA format, and save this file to the local user hard drive. Open the BatchPrimer3 Web page (http://probes.pw.usda.gov/batchprimer3/) and upload the text file using the "Choose File" button. Once completed, continue the primer design process from **step 4** onwards.

General qPCR primer picking parameters applicable to both Primer3 and BatchPrimer3 primer design interfaces are described from **step 4** onwards:

4. Check both the "Pick Left Primer" and "Pick Right Primer" buttons.
5. Adjust preferred amplicon product sizes to between 50 and 150 base-pairs.
6. Adjust primer Tm to 57 °C (minimum), 63 °C (maximum), and 60 °C optimal. (These are the default settings in both Primer3 and BatchPrimer3; *see* **Note 2**).
7. Adjust product Tm to 85 °C (minimum) and 95 °C (maximum).
8. Adjust primer GC% to 40 % (minimum) and 60 % (maximum).
9. Click "Pick Primers" button to generate primers.
10. Primer3 and BatchPrimer3 output primer sequences on a Web interface. To facilitate easy management of designed primers, copy primer sequences into a local database.
11. Undertake a BLASTN search (http://www.ncbi.nlm.nih.gov/blast/Blast.cgi?PAGE=Nucleotides) of the designed primers against the "Nucleotide collection" database in order to ensure their specificity to the desired sequences. BLASTN of multiple primer sequences can be achieved by performing a batch BLASTN search. A batch BLASTN search is carried out by uploading a file containing multiple primer sequences in FASTA format to the BLASTN Web interface using the "Choose File" button. Clicking the BLAST button will execute the search command.
12. Primers for qPCR can be ordered from an appropriate supplier (*see* **Note 3**). Standard PCR primer synthesis conditions are adequate for primers to be used in qPCR.
13. Make up working stocks of oligonucleotide primers by suspending primer pairs in nuclease-free water to a concentration of 1.3 μM each.

3.2 Purification of Total RNA from Bacterial Isolates

This section describes methodologies for the extraction of RNA molecules from pure bacterial cultures. Alternative methods may be required when working with environmental samples and/or non-culturable bacterial strains.

1. Grow 20 mL cultures of target organisms in suitable growth media and glassware/plasticware (*see* **Note 4**) until mid-exponential phase is attained (*see* **Note 5** for descriptions of alternative experimental setups).
2. Harvest cells by centrifugation (5,000×*g*/5 min/4 °C) (*see* **Note 6**). Discard the supernatant and immediately add 1 mL TRIzol® Reagent (Invitrogen) to the cell pellet (*see* **Notes 7** and **8**). Homogenize cells in TRIzol® Reagent (Invitrogen) by triturating with a P1000 pipette (*see* **Note 9**). Transfer cell solution into a 1.5 mL microcentrifuge tube.
3. Extract the aqueous phase with the addition of 200 µL of chloroform (*see* **Note 7**; Sigma-Aldrich; Molecular Biology Grade) to each sample and vortex for 20 s. Centrifuge at 12,000×*g*/15 min/4 °C to separate the phases.
4. Remove the aqueous (clear upper) phase and process the sample using the PureLink™ Micro-to-Midi™ Total RNA Purification System (Invitrogen) following the manufacturer's instructions (*see* **Note 10**). Elute purified RNA in 45 µL RNAse-free water into a sterile 1.5 mL microcentrifuge tube. In order to maximize RNA yield, repeat the elution step with a further 45 µL RNAse-free water. Collect this eluate into the microcentrifuge tube containing the original eluted sample.
5. Add 10 µL of 10× Ambion TURBO™ DNase I Buffer (*see* **Note 11**). Add 5 µL Ambion TURBO™ DNase I to the buffered RNA solution incubate reaction tubes at 37 °C for 60 min to degrade any contaminating DNA species. Inactivate the DNase I enzyme by heating the sample to 75 °C for 10 min.
6. Clean up the DNase I treated RNA sample using the PureLink™ Micro-to-Midi™ Total RNA Purification System (Invitrogen). Elute the DNase I treated sample in 30 µL nuclease-free water.
7. Determine the quantity and quality of RNA using a NanoDrop™ spectrophotometer (*see* **Note 12**). Store RNA aliquots at −80 °C until required.

3.3 Reverse-Transcription Reaction in the Detection of RNA Molecules

1. For each sample, dilute 2.5 µg of purified RNA into a total volume of 14 µL. Add 4 µL of 5× VILO™ Reaction Mix (Invitrogen) and 2 µL of 10× SuperScript® Enzyme Mix (Invitrogen) to the each reaction (*see* **Note 13**).
2. For synthesis of cDNA species, incubate samples at 25 °C for 5 min, then at 42 °C for 30–90 min, and finally, at 85 °C for 5 min, to inactivate the enzyme (*see* **Note 14**).

3. Dilute the cDNA samples to a concentration equivalent to 5 ng/μL of starting RNA in a sterile 1.5 mL tube, e.g., add 480 μL RNAse-free water to 20 μL samples that included 2.5 μg total RNA.
4. "No-RT" (negative reverse transcription) control reactions must be conducted to detect undigested genomic DNA in cDNA preparations. To prepare no-RT controls, set up the reverse transcription reactions according to **step 1** of this section, but omit the reverse transcriptase enzyme from reaction tubes (substitute 2 μL RNase-free water for the RT enzyme). Conduct **steps 2** and **3** (this section) and use the product in control reactions (*see* **Note 15**).
5. Store cDNA aliquots and no-RT controls at −20 °C until required.

3.4 Purification of Genomic DNA from Bacterial Isolates

This section describes methodologies for the extraction of genomic DNA from pure bacterial cultures. Alternative methods may be required when working with genomic DNA to be extracted from environmental samples and/or non-culturable bacterial strains.

1. Grow pure bacterial cultures according to **step 1** of Subheading 3.2. Transfer a 1 mL aliquot from the mid-exponential phase culture to a sterile 1.5 mL microcentrifuge tube. Pellet bacterial cells by centrifugation at 10,000×*g* for 1 min.
2. Remove the supernatant from the bacterial cell pellet and process the sample using the ISOLATE Genomic DNA mini-kit (Bioline Reagents Limited, USA) following the manufacturer's instructions (*see* **Note 16**). Elute purified genomic DNA in 200 μL sterile water in a 1.5 mL microcentrifuge tube and dilute to a concentration of 5 ng/μL for use in qPCR. Store at 4 °C until required.

3.5 qPCR Reaction Setups for the Quantification of Genetic Markers from cDNA or gDNA

This section describes generalized methodologies for the setup and real-time thermal cycling of qPCR reactions using synthesized cDNA or isolated gDNA as reaction template molecules. Step-by-step methods for the quantification of differential gene expression within cDNA populations, and for the detection of gene copies in environmental samples using isolated gDNA, will be described in sections which follow.

1. Thaw 2× GoTaq® qPCR Master Mix on ice (*see* **Note 17**) and briefly centrifuge the reagent tube at low speed in order to collect the Master Mix to the bottom of the tube. Store Master Mix on ice until required.
2. Set up reaction mixes in appropriate plasticware (e.g., white well tubes in the MasterCycler® ep *realplex*; Eppendorf International, Hamburg, Germany) according to Table 1, in the order presented. Precise addition of the components will result in a total qPCR reaction volume of 5 μL. (For further information

Table 1
Typical qPCR components

Component[a]	Volume per 5 μL reaction	Final concentration
2× GoTaq® qPCR Master Mix	2.5 μL	1×
Primer mix	1.5 μL	0.4 μM

[a]Adapted from Promega's GoTaq® qPCR Master Mix product insert

Table 2
Typical cycling parameters for qPCR using GoTaq

	Number of cycles[a,b]	Cycling program
Hot Start activation	1	95 °C for 2 min
Denaturation	1	95 °C for 15 s
Annealing/extension	40	60 °C for 60 s

[a]Adapted from Promega's GoTaq® qPCR Master Mix product insert
[b]The protocol presented represents a two-step cycling method, where the annealing and extension steps are combined. Alternative three-step protocols may be required for optimization of qPCR, especially where the primer Tm values significantly differ from the values suggested in Subheading 3.1

regarding the setup of qPCR reactions, *see* **Note 18**). Since qPCR volumes outlined in this protocol are relatively small, it is important that pipettes are calibrated and pipetting manipulations are accurate.

3. Add 1 μL of cDNA (Subheading 3.4) or gDNA (Subheading 3.5) to the reaction mixes.
4. Set up an appropriate number of "no-template" control reactions. "No-template" control reactions enable an assessment of contamination in individual reaction components (qPCR Master Mix or Primer Stock). To set up "no-template" controls, replace the cDNA or gDNA templates in the reaction setups above with the equivalent volume of water. Ideally, no amplification should be observed in these control reactions.
5. Seal reaction tubes and gently flick to mix contents. Collect qPCR reaction mixes to the bottom of the PCR tubes by low-speed centrifugation.
6. A thermal cycling regime suitable for reactions containing GoTaq® qPCR Master Mix is outlined in Table 2. Program this cycling regime into a suitable thermal cycler with optical detection module fitted (*see* **Note 19**). Programming protocols are specific for each thermal cycler; consult the equipment manual for specific information regarding thermal cycle programming.

Table 3
Typical qPCR thermal cycling regime with melt curve analysis included

	Number of cycles[a,b]	Cycling program
Hot Start activation	1	95 °C for 2 min
Denaturation	1	95 °C for 15 s
Annealing/extension	40	60 °C for 60 s
Melt curve analysis	1	60–95 °C over 20 min

[a]Table is adapted from Promega's GoTaq® qPCR Master Mix product insert
[b]The protocol presented represents a two-step cycling method, where the annealing and extension steps are combined. Alternative three-step protocols may be required for optimization of qPCR, especially where the primer Tm values significantly differ from the values suggested in Subheading 3.1

7. GoTaq® qPCR Master Mix contains a patented DNA-binding dye which exhibits similar spectral properties to SYBR® Green I. For real-time detection of DNA amplification, use the qPCR machine software to set the optical detection module to read SYBR® Green I or FAM™ fluorescence. GoTaq® qPCR Master Mix also contains CRX as a reference dye, which has the same spectral properties as the commonly used reference dye, ROX™ (*see* **Note 20**). Set the detection module of the real-time PCR machine to detect ROX™ for all reactions to be completed.
8. Load the qPCR reaction tubes into the thermal cycler, and start the thermal cycling program.

3.6 Evaluation of PCR Efficiency

Evaluation of PCR efficiency is an important initial consideration in the establishment of an effective qPCR regime. Inefficiencies in PCR may indicate suboptimal primer–template binding or reduced DNA polymerase activity, whilst theoretical PCR efficiency calculations of over 100 % indicate the presence primer dimers in the reaction or incorrect pipette calibration. Assessment of PCR efficiency should be conducted prior to performing experimental assays on synthesized cDNA or on sample extracted genomic DNA, since suboptimal qPCR may introduce experimental error into the quantification of test nucleic acids.

1. In sterile, nuclease-free plasticware, make five 10-fold dilutions of test cDNA or gDNA.
2. Using these serially diluted samples, set up qPCR reactions according to the methodology outlined in Subheading 3.5. Ensure that no-template qPCR controls are included.
3. Program the thermal cycler as per Subheading 3.5, with an additional cycling step to conduct a melt (dissociation) curve analysis (Table 3). Melt curve analyses provide an assessment

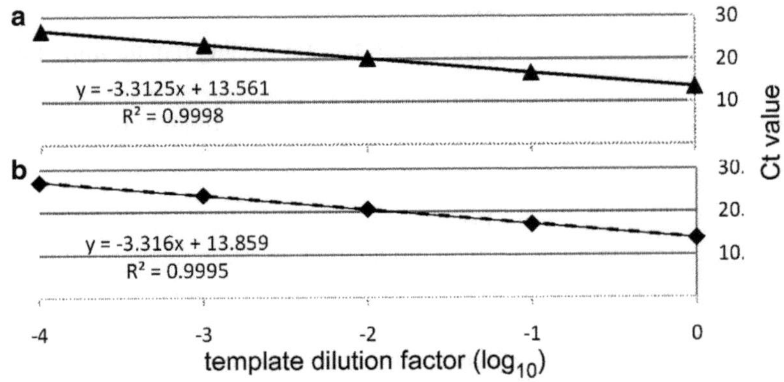

Fig. 2 Amplification efficiency plots for primer sets amplifying *GAPDH* and *aedD* gene fragments from *Acinetobacter baumannii* cDNA. Amplication efficiencies calculated from the gradient of the line of best fit (using Eq. 1, Subheading 3.6) are 100.4 % (a) and 100.2 % (b)

of the specificity of the PCR for the target DNA; a specific PCR will produce a single product at high temperature, as indicated by a single rapid drop in fluorescence, or a single peak on the dt/dN melt curve. An option for melt curve analysis is a feature of the software of most qPCR machines.

4. Start the thermal cycling regime.
5. Once qPCR is complete, download and save C_T (*see* **Note 21**) values into a spreadsheet database.
6. Using the spreadsheet software, create an XY scatter plot of the C_T values versus the log of the cDNA or gDNA dilution and once completed, perform a linear analysis of the data points (Fig. 2). Ensure that the correlation coefficient of the line (R^2 value) is greater than 0.985.
7. Use Eq. 1 to determine the efficiency of the PCR from the slope of the linear regression line constructed in **step 6**. Using Eq. 1 the efficiency of the PCR shown in Fig. 2a is 100.4 % and that of the PCR in Fig. 2b is 100.2 %.

$$\text{Efficiency} = 10^{(-1/\text{slope})} - 1 \times 100 \tag{1}$$

8. Calculated PCR efficiencies of between 80 and 110 % are considered to be in the acceptable range for qPCR analysis. Should the calculated efficiency fall outside this range, redesign or re-optimize the experiment (*see* **Note 22**).

3.7 Quantification of mRNA Expression Using the Comparative C_T ($\Delta\Delta C_T$) Method

The $\Delta\Delta C_T$ method of quantification expresses the abundance of an mRNA in a test sample population relative to that of the abundance of the same mRNA species in a control or baseline sample [9]. This method assumes amplification efficiencies close to 100 %

(*see* Subheading 3.6). If such efficiencies cannot be achieved after qPCR optimization, alternative calculations are required (e.g., [10]). The experimental C_T values are normalized against an endogenous control gene, which should be constitutively expressed in all samples tested, and unaffected by any test conditions applied. Described mathematically, this summation is expressed using the formula in Eq. 2.

$$Relative\ abundance = 2^{-\Delta\Delta C_T} \tag{2}$$

This section describes the calculation of the relative abundance of mRNA molecules in a test population using the $\Delta\Delta C_T$ relative method of quantification (*see* **Note 23**).

1. Set up reactions for real-time thermal cycling as per Subheading 3.5.
2. Upon completion of the cycling regime, use the thermal cycler software to import C_T values for all samples into a spreadsheet database.
3. Using the spreadsheet software, calculate the ΔC_T values for both the test sample and the control sample, by employing the formulae in Eqs. 3 and 4 (Fig. 1).

$$\Delta C_{T(\text{test sample})} = C_{T(\text{target cDNA})} - C_{T(\text{endogenous reference cDNA})} \tag{3}$$

$$\Delta C_{T(\text{baseline sample})} = C_{T(\text{target cDNA})} - C_{T(\text{endogenous reference cDNA})} \tag{4}$$

4. Calculate the $\Delta\Delta C_T$ values for the test samples using Eq. 5.

$$\Delta\Delta C_T = \Delta C_{T(\text{test sample})} - \Delta C_{T(\text{baseline sample})} \tag{5}$$

5. To quantify the relative abundance of the target cDNA in the test sample normalized to the endogenous reference cDNA, apply the formula for relative abundance (Eq. 2).

3.8 Quantification of Target Gene Copies in a Sample Using a Standard Curve

Quantification of the number of copies of a target gene in a particular sample can be achieved with the use of a standard curve. If the copies of target gene per genome are known, standard-curve based qPCR analysis can be used to assess the total number of gene copies, and therefore organisms, in a particular cross section of a sampled population. Since the standard-curve methodology described here is not a comparative means of quantification, an endogenous control gene is not required. However, to construct the standard curve, a purified DNA sample containing a known number of copies of the target gene is necessary. DNA standards used in this assay may be genomic DNA, plasmid DNA, or a PCR product. For an assessment of the number of copies in a particular test sample to be made, the number of copies of the gene target per microlitre of DNA standards must be known (*see* **Note 24**).

1. In fresh tubes, make five 10-fold dilutions of the DNA standard. Set up qPCR reactions and commence thermal cycling as per Subheading 3.5.
2. Upon completion of the cycling regime, use the thermal cycler software to import C_T values for all samples into a spreadsheet database.
3. Using the spreadsheet software, construct an XY scatter plot of the C_T values versus the log of the DNA standard, and, once completed, perform a linear regression analysis of the data points. Ensure that the correlation coefficient of the line (R^2 value) is greater than 0.985.
4. Use the spreadsheet software to calculate the equation of the linear regression line in the form:

$$y = mx + b \tag{6}$$

In this standard-curve based copy analysis, y= the C_T value and x= log of the copies of the DNA standard, so the equation can be rearranged:

$$N = 10^{\frac{(C_T - b)}{m}} \tag{7}$$

where N= the number of DNA copies in the sample analyzed.

5. Extract gDNA from samples as per Subheading 3.4 and conduct qPCR on the isolated gDNA as per Subheading 3.5.
6. Import the C_T data for the test samples into the spreadsheet database and determine the number of copies of the target gene in the DNA sample by applying Eq. 7.

4 Notes

1. Commonly used endogenous control genes for qPCR analyses include *rpoB*, *gyrB*, *recA*, 16S rRNA, and *ihfB*. Other reference genes may be used, but these genes must suit two essential criteria: (1) reference genes must be constitutively expressed in the organism under study, ideally at a level similar to the experimental gene(s) under investigation, and (2) the expression of reference genes must not be influenced by changes in growth or treatment conditions.
2. Selection of appropriate Tm and GC ranges is dependent on the GC content of the organism under study. The suggested Tm and GC ranges presented are relevant to organisms containing an average GC % of between 30 and 70. Tm and GC parameter values may be reduced for low CG content organisms or increased for high GC content organisms. Some optimization of cycling parameters may be required if these values significantly differ from the suggested ranges.

3. For ease of use when working with numerous primers, it may be appropriate to order primers from the supplier in a microplate format. Similar to primers ordered in tube formats, primer microplates are generally supplied to the customer containing purified lyophilised oligonucleotides. Some primer suppliers offer the additional option of supplying primers prediluted in water or TE, to a concentration suitable for the user's application. Additionally, various oligonucleotide supply companies offer the option of creating a "forward" and "reverse" primer stock mix when ordering primers in the microplate format. This option may be convenient for users desiring minimal manipulation steps, for example, when qPCR reactions are being set up by hand. Contact the supplier for further information regarding these alternative format options.
4. Plasticware and/or glassware used in cell culturing/collection should be sterile and nuclease free. RNases may be removed from laboratory glassware by soaking items in a solution of 0.1 % diethylpyrocarbonate (DEPC) overnight. Autoclave DEPC treated glassware (121 °C, 15 min) in order to degrade residual DEPC prior to use. Work surfaces and items unable to be treated with DEPC solutions may be decontaminated from RNases with a commercial preparation, such as *RNAseZAP*® (Ambion, USA).
5. At least three independent biological replicates should be used for each condition. Alternative experimental setups testing the effects of specific environmental factors on transcript expression in a single strain may be required. Depending on the application and the genes under investigation, mid-exponential phase may not be the most suitable point in cellular growth to harvest cells and this point should be determined empirically. Organisms may be isolated directly from environmental sources for qPCR, and in these cases laboratory culture of cells may not be required. When laboratory culture of target organisms is necessary, ensure that culture conditions are suited to the organism under study (e.g., temperature, growth media, aerobic/anaerobic conditions).
6. If total RNA extraction is unable to be conducted immediately, flash-freeze harvested cells using liquid nitrogen and store at −80 °C.
7. TRIzol® Reagent contains phenol and guanidium thiocyanate, which are potentially hazardous. Chloroform is a toxin and a potential carcinogen. Consult product Material Safety Data Sheets (MSDS) for further information.
8. A product such as RNA*later* (Ambion, USA) may be added to cells prior to centrifugation in order to preserve RNA, e.g., if studies into RNA half-life are to be conducted or the RNA cannot be extracted immediately and flash-freezing is not an option.

9. Suspending cellular material in TRIzol® Reagent (Invitrogen) will facilitate cell lysis, prevent RNA degradation from endogenous RNases, and act as initial step for protein extraction. Some strains require enzymatic treatments or physical abrasion in order to breach outer cell walls. If this step is relevant to the target organism, ensure that this step is carried out prior to TRIzol® Reagent treatment.
10. We have successfully used the Invitrogen PureLink Micro-to-Midi™ RNA extraction kit for our RNA purification procedures prior to qPCR. However, several high-quality commercial RNA extraction kits are available from alternative suppliers (*see* Subheading 2) that may be better suited to individual needs and budgets. Alternatively, excellent quality RNA may be extracted from many bacterial species using standard protocols, particularly following the rapid total RNA isolation method of Chomczynski and Sacchi (1987) [11].
11. We find post-purification DNase I treatment of RNA preparations results in a greater efficiency of DNA digestion than on-column methodologies. However, should the user require an on-column DNAse I treatment, this step may easily be incorporated into the purification protocol for the Invitrogen Micro-to-Midi™ Total RNA Purification System. Alternative commercially available kits which employ a column purification step for the isolation of RNA molecules should also be equally amenable to on-column DNase I treatment of sample preparations.
12. The NanoDrop™ spectrophotometer provides a rapid and efficient platform for the quantification of nucleic acids, and is the preferred system in our laboratory. However, if this system is unavailable in the user's laboratory, please take note of the following considerations for the determination of RNA concentration and purity:

    The absorbance maxima of single stranded RNA (ssRNA) is 260 nm. This property can be used to determine the concentration of RNA in a purified sample. The Beer–Lambert Law relates absorbance to the RNA concentration of a given sample using the following equation:

    $$A = \varepsilon C l$$

    where A = absorbance of the sample at 260 nm, ε = absorption coefficient of ssRNA, C = concentration of the sample, l = pathlength of the spectrophotometer cuvette

    The average absorption coefficient of ssRNA is 0.025/(μg/mL) cm. Assuming a cuvette pathlength of 1 cm (typical for most laboratory spectrophotometers), an A_{260} of 1 equates to a ssRNA concentration of 40 μg/mL. The concentration of ssRNA (in μg/mL) can therefore be calculated using the equation:

$$\text{Concentration of ssRNA}\left(\mu g / mL\right) = A_{260} \times \left(\text{dilution factor applied}\right) \times 40\ \mu g / mL$$

For downstream applications using ssRNA (e.g., reverse transcription), it is important that the RNA preparation is free from proteins and other contaminants. Proteins absorb maximally at 280 nm, and the $A_{260/280}$ ratio can be used to calculate the level of protein contamination in a sample preparation. A ssRNA preparation completely free of contaminating protein species has an $A_{260/280}$ ratio of 2. Organic solvents and common RNA purification reagents, such as guanidium thiocyanate and phenol, will absorb at around 230 nm. Therefore, the $A_{260/230}$ ratio provides a measure of solvent–reagent contamination in RNA samples. The $A_{260/230}$ ratio of a pure RNA sample is considered to be >2.

13. If less than 2.5 μg of RNA is available, or the concentration of the RNA sample is <178.5 ng/μL (i.e., the minimum concentration in 14 μL which provides a reaction amount of 2.5 μg total RNA), a smaller amount of RNA can be used. However, ensure that amounts of RNA to be used for reverse transcription are standardized across each reaction. A range of alternative cDNA synthesis reagents are available from alternative suppliers (*see* Subheading 2).

14. This cycling regime can be conveniently carried out with the use of a thermal cycler.

15. No-RT control reactions should be set up as described in Subheading 3.5, except that the no-RT template is substituted for the cDNA template. The cycle threshold (C_T) for the no-RT control reaction should be at least five cycles greater than the C_T value for the corresponding cDNA samples. Using off column DNase digestions as described in Subheading 3.2, **step 5**, products are generally not detected in no-RT reactions below 40 cycles of PCR.

16. Many commercially available kits may be adapted for the extraction of genomic DNA from environmental bacterial species/samples, and we routinely use the ISOLATE Genomic DNA mini-kit manufactured by Bioline Reagents Limited, USA, for this purpose. As with other commercially available kits, we recommend that the reader research alternative options in order to ascertain kits suitable for their individual needs and budgets. Standard isolation protocols post cell-lysis (for example, *see* ref. 12) may be employed in the production of high quality genomic DNA for qPCR.

17. We have achieved good real-time PCR results using the 2× GoTaq® qPCR Master Mix available from Promega. Numerous reagent supply companies also stock their own lines of qPCR Master Mixes and other reagents useful in qPCR. The reader is encouraged to research alternative suppliers in order to ascertain appropriate products which suit individual requirements and budgets. 2× GoTaq® qPCR Master Mix contains a proprietary DNA binding dye (known as BRYT GREEN™), which has spectral properties similar to those of SYBR® Green I. Fluorescent DNA binding dyes are light-sensitive, and as such exposure to light should be kept to a minimum. We find that wrapping reagent tubes containing fluorescent dyes using aluminum foil is an effective way of reducing the exposure of dyes to light during reagent manipulations.
18. 5 μL reaction volumes are the minimum size which can be applied in qPCR. Reactions may be scaled up appropriately if desired. If numerous qPCRs are to be set up, a liquid-handling robot (such as the epMotion®, Eppendorf International, Hamburg, Germany) is recommended.
19. We use a MasterCycler® ep *realplex* (Eppendorf International, Hamburg, Germany) for qPCR analyses. Numerous suppliers of real-time PCR machinery are available, and the reader is encouraged to research alternative options in order to ascertain equipment which is suitable to their specific needs and budget.
20. Many Applied Biosystems instruments require the addition of carboxy-X-rhodamine (CRX) to qPCR reactions prior to thermal cycling (check individual instrument details to determine if this applies). The GoTaq® qPCR Master Mix is supplied with a 100× solution of CRX; simply add 100× CRX to a concentration of 1× in qPCR reactions if needed. Total qPCR reaction volumes may be required to be increased in order to accommodate the additional CRX reaction component.
21. The C_T (Threshold Cycle) values is the cycle at which the fluorescence signal becomes detectable by the optical detection module of the real-time thermal cycler above the background fluorescence level (Fig. 1). The C_T values for all samples are recorded and saved by the thermal cycler software, and can then be imported into a spreadsheet database for ease of analysis.
22. Causes of low PCR efficiency include:
    (a) *Poor primer design*. Redesign primers in a location shifted away from the original design site.
    (b) *Reduced activity of DNA polymerase in the qPCR Master Mix*. Use a fresh aliquot of Master Mix in subsequent qPCR reactions.

(c) *Secondary structure of the amplified product.* Design primers to amplify a product distal to the original product, check for possible secondary structure in the new region.

Causes of PCR efficiency over 110 %:

(a) *Inaccurate pipette calibration.* Recalibrate pipettes and ensure that pipetting manipulations are accurate.

(b) *Prime dimer.* Reduce the concentration of primers in the reaction or increase the annealing temperature of the thermal cycling regime.

23. When employing the $\Delta\Delta C_T$ method of relative quantification, the amplification efficiencies of the target cDNA and the endogenous reference cDNA should be approximately equal. To determine the amplification efficiency of a particular reaction, refer to Subheading 3.6. **Note 22** outlines methodologies for improving PCR efficiency.

24. What follows is a worked example for the calculation of DNA copies per µL of purified DNA. In this example, the *E. coli* genome, which consists of 4.64×10^6 base pairs, will be used. Assuming the average mass of a nucleotide base pair is 660 Da, the molecular weight of the *E. coli* genome is:

$$660 \times 4.64 \times 10^6 = 3 \times 10^9 \, \text{Da (approx.)}$$

Therefore, one mole of *E. coli* genomic DNA $= 3 \times 10^9$ g

Using Avogadro's constant, we know that 3×10^9 g of *E. coli* genomic DNA contains 6.022×10^{23} copies of a single copy gene. Therefore, the amount of *E. coli* genomic DNA containing a copy of single copy gene is:

$$\frac{3 \times 10^9}{6.022 \times 10^{23}} = 4.98 \times 10^{-15} \, \text{g}$$

Therefore, 4.98 femtograms of *E. coli* genomic DNA contains one copy of a single copy gene. One microlitre of a solution of purified *E. coli* genomic DNA at a concentration of 1 µg/µL would therefore contain 1×10^9 copies of a single copy gene.

Acknowledgements

Part of this work was supported by Project Grant 535053 from the National Health and Medical Research Council. KH is supported by an Australian Postdoctoral Fellowship funded by the Australian Research Council (grant DP110102680).

References

1. Bustin SA, Benes V, Garson JA, Hellemans J, Huggett J, Kubista M, Mueller R, Nolan T, Pfaffl MW, Shipley GL, Vandesompele J, Wittwer CT (2009) The MIQE guidelines: minimum information for publication of quantitative real-time PCR experiments. Clin Chem 55:611–622
2. VanGuilder HD, Vrana KE, Freeman WM (2008) Twenty-five years of quantitative PCR for gene expression analysis. Biotechniques 44:619–626
3. Higuchi R, Fockler C, Dollinger G, Watson R (1993) Kinetic PCR analysis: real-time monitoring of DNA amplification reactions. Biotechnology 11:1026–1030
4. Holland PM, Abramson RD, Watson R, Gelfand DH (1991) Detection of specific polymerase chain reaction product by utilizing the 5′–3′ exonuclease activity of Thermus aquaticus DNA polymerase. Proc Natl Acad Sci U S A 88:7276–7280
5. Tyagi S, Kramer FR (1996) Molecular beacons: probes that fluoresce upon hybridization. Nat Biotechnol 14:303–308
6. Rozen S, Skaletsky H (2000) Primer3 on the WWW for general users and for biologist programmers. Methods Mol Biol 132:365–386
7. You FM, Huo N, Gu YQ, Luo MC, Ma Y, Hane D, Lazo GR, Dvorak J, Anderson OD (2008) BatchPrimer3: a high throughput web application for PCR and sequencing primer design. BMC Bioinformatics 9:253
8. Brzoska AJ, Hassan KA, de Leon EJ, Paulsen IT, Lewis PJ (2013) Single-step selection of drug resistant Acinetobacter baylyi ADP1 mutants reveals a functional redundancy in the recruitment of multidrug efflux systems. PLoS ONE 8:e56090
9. Livak KJ, Schmittgen TD (2001) Analysis of relative gene expression data using real-time quantitative PCR and the 2(-Delta Delta C(T)) Method. Methods 25:402–408
10. Pfaffl MW (2001) A new mathematical model for relative quantification in real-time RT-PCR. Nucleic Acids Res 29:e45
11. Chomczynski P, Sacchi N (1987) Single-step method of RNA isolation by acid guanidinium thiocyanate-phenol-chloroform extraction. Anal Biochem 162:156–159
12. Sambrook J, Russell DW (2001) Molecular cloning: a laboratory manual, 3rd edn. Cold Spring Harbor Laboratory, Cold Spring Harbor, NY
13. Hassan KA, Brzoska AJ, Wilson NL, Eijkelkamp BA, Brown MH, Paulsen IT (2011) Roles of DHA2 family transporters in drug resistance and iron homeostasis in Acinetobacter spp. J Mol Microbiol Biotechnol 20:116–124

Part II

Describing Microbial Communities

Chapter 4

Analysis of Community Dynamics in Environmental Samples Using Denaturing Gradient Gel Electrophoresis

Claire L. Thompson

Abstract

Denaturing gradient gel electrophoresis (DGGE) is a culture-independent fingerprinting technique that allows for rapid comparative analysis of changes to microbial communities. 16S rRNA genes amplified from environmental samples can be separated based on their melting behavior in a denaturing gradient of urea and formamide. A fingerprint of the microbial community is generated with each band on the gel assumed to correspond to a different bacterial species. Community dynamics can then be assessed through statistical analysis of DGGE profiles and the sequencing of excised bands.

Key words Denaturing gradient gel electrophoresis, DGGE, 16S rRNA gene, Bacterial communities, Fingerprinting, Community dynamics, Diversity

1 Introduction

Microbial communities are highly dynamic systems which can exhibit both spatial and temporal variation and are also capable of responding to environmental stress. Understanding the dynamics of a community is an important step in being able to make accurate predictions regarding community behavior or to engineer particular outcomes. However, the ability to monitor changes in the composition of microbial communities requires a technique that can rapidly compare a series of samples. While approaches such as the sequencing of cloned genes provide detailed phylogenetic information, they are impractical for comparing a large number of samples simultaneously.

Community fingerprinting techniques including denaturing gradient gel electrophoresis (DGGE) are based on the rapid analysis of marker genes recovered by PCR (often 16S rRNA) and have proven to be valuable tools for rapidly generating profiles of diversity in order to characterize a range of microbial communities [1]. While this technique is considered to detect microorganisms above 1 % in abundance [1], populations of lower abundance can be

Ian T. Paulsen and Andrew J. Holmes (eds.), *Environmental Microbiology: Methods and Protocols*, Methods in Molecular Biology, vol. 1096, DOI 10.1007/978-1-62703-712-9_4, © Springer Science+Business Media, LLC 2014

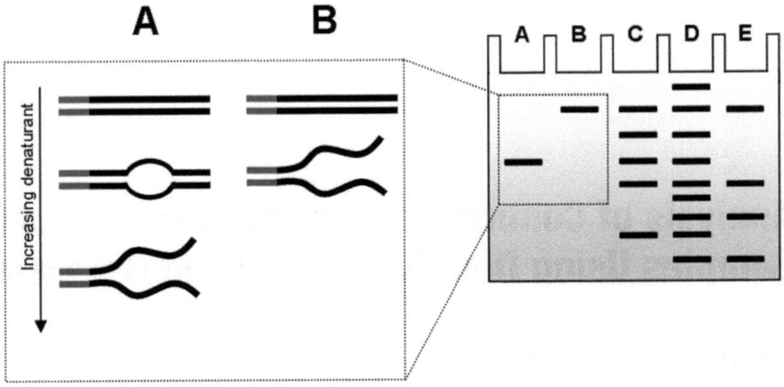

Fig. 1 Denaturing gradient gel electrophoresis. Migration of DNA in the gel is determined by its melting behavior in a gradient of urea and formamide. DNA fragments of the same size are separated, generating a fingerprint of community diversity

investigated using group-specific DGGE analyses either by a nested PCR approach [2, 3] or using group-specific primers that are compatible with DGGE [4, 5].

DGGE involves the separation of double-stranded DNA fragments that are of the same size but which differ in sequence (Fig. 1). The mobility of the fragments depends on their melting behavior in a denaturing gradient of urea and formamide. As fragments migrate further along the polyacrylamide gel they encounter increasingly higher concentrations of denaturants. Sequence differences cause the fragments to melt at different positions along the gradient. Partially denatured fragments will migrate more slowly than non-denatured fragments. In order to prevent the two DNA strands from completely dissociating and to detect mutations in the higher melting domains, a GC-rich fragment (GC-clamp) is added to one of the PCR primers. As a result, different sequences will migrate different distances along the gel, generating a DGGE profile that is characteristic of the microbial community. When interpreting DGGE profiles, it is assumed that each band corresponds to a different bacterial species. However it should be kept in mind that some bands may be comprised of more than one co-migrating species or that some species may contribute more than one band if they possess multiple different copies of the 16S rRNA gene.

Multiple DGGE profiles can be compared in order to observe spatial or temporal differences within a microbial community (Fig. 2). This technique has previously been used for studying temporal changes occurring in the intestinal microbiota of mammals [6–8] as well as in soil communities [9]. One of the major advantages of DGGE when compared with other fingerprinting techniques such as terminal restriction fragment length polymorphism (T-RFLP) is that the identity of bands in the gel can be determined through excision and sequencing. Therefore, it is possible to determine the identity of bacteria that are involved in changes to community composition.

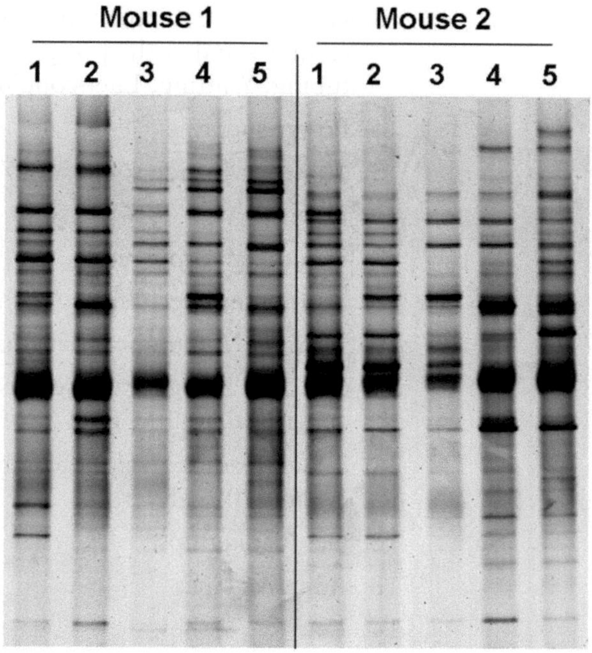

Fig. 2 DGGE analysis of temporal changes in the gut community of mice. DGGE profiles were generated from fecal samples collected on a daily basis from two C57BL/6 mice deficient in the Interferon regulatory factor 9

2 Materials

2.1 PCR

1. Environmental DNA (*see* **Note 1**).
2. PCR Thermocycler.
3. 10× ThermoPol reaction buffer (New England Biolabs, Beverly, MA, USA).
4. Deoxynucleotide solution mix (dNTPs) (5 mM).
5. F-968-GC and 1401r PCR primers (0.05 mM) (*see* **Note 2**).
6. Taq DNA Polymerase (5 U/μl) (New England Biolabs).
7. PCR-grade water.

2.2 DGGE

A number of different DGGE systems are available such as the INGENYPhorU (Ingeny, Leiden, The Netherlands) and DGGE-2001 systems (CBS Scientific, Solana Beach, CA, USA). The protocol here describes the use of the DCode System (Bio-Rad, Hercules, CA, USA) for casting 16 cm gels.

1. DGGE apparatus including electrophoresis/temperature control module, electrophoresis tank, and sandwich core.
2. 16 cm gel set including 2 glass plates, 2 sandwich clamps, 1 mm spacers, and 1 mm 16-well comb.

3. Gel casting equipment including Model 475 gradient former, casting stand with sponges, alignment card, 30 ml syringes, tubing, Y-fitting, Luer couplings, and syringe locks.
4. Power supply.
5. 40 % Acrylamide solution: 37.5:1 acrylamide–bis-acrylamide.
6. Urea.
7. Deionized formamide.
8. Ammonium persulfate (APS).
9. *N*,*N*,*N*′,*N*′-tetramethylenediamine (TEMED).
10. 50× TAE: Dissolve 242 g Tris base, 57.1 ml glacial acetic acid, and 100 ml 0.5 M EDTA (pH 8) in 1 L distilled water.
11. 2× gel-loading dye: 0.05 % bromophenol blue, 0.05 % xylene cyanol, 70 % glycerol.
12. Gel-loading tips.
13. Adhesive tape.
14. Lint-free tissues.

2.3 Gel Staining

All solutions should be made with distilled water.

1. 10 % ethanol solution.
2. 1 % nitric acid solution.
3. 0.2 % silver nitrate solution.
4. Developer solution: 29.2 g/L sodium carbonate, 0.05 % formalin.
5. 3 % acetic acid.

2.4 Sequencing of DGGE Bands

1. Sterile scalpels.
2. PCR-grade water.

2.5 Image Analysis

1. Gel image capture system, e.g., GS-800 Densitometer (Bio-Rad).
2. Image analysis software (*refer to* Subheading 3.5).

3 Methods

3.1 PCR

1. Prepare a mastermix with the components found in Table 1.
2. Aliquot 24 μl of mastermix into each tube and add 10–100 ng of environmental DNA.

Table 1
PCR mastermix components

10× ThermoPol buffer	2.5 μl
dNTPs (5 mM)	1.0 μl
F-968-GC (0.05 mM)	0.4 μl
R-1401 (0.05 mM)	0.2 μl
Taq DNA polymerase (5 U/μl)	0.2 μl
PCR-grade water	19.7 μl

3. Place the PCR tubes in the thermocycler and start the following program: 1 min of initial denaturation at 94 °C, followed by 30 cycles of denaturation (94 °C for 30 s), annealing (56 °C for 30 s), and extension (72 °C for 60 s) with a final extension for 7 min at 72 °C.
4. PCR products should be assessed for quality and yield by agarose gel electrophoresis prior to DGGE analysis.

3.2 DGGE

3.2.1 Casting the Gel

1. Clean and dry glass plates and spacers using 95 % ethanol and lint-free tissues. Flush all tubing and syringes with water (*see* **Note 3**).
2. Form the gel sandwich by placing the spacers along the shorter edges of the large glass plate and then place the smaller glass plate on top.
3. Attach the sandwich clamps to either side of the plates and tighten the screws so that the plates are held in place.
4. Put the gel sandwich into the alignment slot of the casting stand. Loosen the sandwich clamps slightly. Ensure that plates and spacers are flush and in alignment using the alignment card before re-tightening the sandwich clamps (*see* **Note 4**).
5. Place the gel sandwich in the casting slot of the casting stand and secure by turning the handles on either side of the casting stand (*see* **Note 4**).
6. Place needle (with the short piece of tubing and Y-fitting attached) in gap between glass plates. Secure with adhesive tape.
7. Prepare 10 % APS solution (*see* **Note 5**).
8. Prepare 25 ml each of the 0 % and 100 % denaturant stock solutions (*see* **Note 6**) (*see* Table 2).
9. Prepare low and high denaturant solutions for a 40–70 % gradient (*see* Table 3) from the 0 % and 100 % stock solutions (*see* **Notes 5** and **6**). APS and TEMED should be added last as gel will set within approximately 10 min after they are added.

Table 2
0 % and 100 % denaturant stock solutions

Components	0 %	100 %
40 % Acrylamide solution	5 ml	5 ml
50× TAE	0.5 ml	0.5 ml
Deionized formamide	–	10 ml
Urea	–	10.5 g
Distilled water	Up to 25 ml	Up to 25 ml

Table 3
Denaturant solutions for a 40–70 % gradient

Components	40 %	70 %
0 % denaturing solution	10.8 ml	5.4 ml
100 % denaturing solution	7.2 ml	12.6 ml
10 % APS	162 μl	162 μl
TEMED	16.2 μl	16.2 μl

10. Fill a syringe (with tubing attached) with slightly more than 15 ml of each acrylamide solution. Remove any bubbles and make sure that solution reaches the end of the tubing.
11. Ensure that gradient maker is set to deliver the correct volume (a setting of 14.5 for 16 cm × 16 cm gels). Load the syringe with 40 % denaturant solution into the sleeve on the gradient maker labelled "low" and the syringe with the 70 % denaturant solution into the sleeve labelled "high." Before pouring the gradient, secure each syringe firmly in the syringe holder and fit the plunger end of the syringe onto lever of the cam wheel.
12. Slowly turn the cam wheel to cast gel. Insert comb when complete.
13. Clean the tubing with water immediately to prevent it from polymerizing in the tubing.
14. Allow gel solution to polymerize for at least 60 min at room temperature.

3.2.2 Running the Gel

1. Fill electrophoresis tank with 7 L 1× TAE (140 ml 50× TAE, 6.85 L water).
2. Place temperature control module on top of the tank

3. Switch on the power, pump, and heater.
4. Set temperature to 60 °C.
5. Allow 1× TAE buffer in the tank to reach 60 °C (requires at least 90 min).
6. Remove the comb from gel and rinse the wells with 1× TAE to remove any unpolymerized acrylamide.
7. Attach the gel sandwich to the sandwich core. If only one gel is to be run, then attach a second empty gel sandwich (with plates and clamps but no spacers or gel) to the other side of the sandwich core (*see* **Note 7**).
8. Switch off the temperature control module and remove approximately 350 ml of 1× TAE from the tank. Put the sandwich core in tank and fill the upper compartment with 350 ml of 1× TAE. Replace lid and turn the heater and pump back on (*see* **Note 8**).
9. Prepare the samples by mixing each PCR product with the gel loading dye. The amount of PCR product loaded onto the DGGE gel depends on the expected complexity of the community (*see* **Note 9**).
10. After the temperature reaches 60 °C, flush wells with 1× TAE from tank and load each sample.
11. Run gel overnight at 80 V for 16 h.

3.3 Gel Staining

There are multiple methods to stain DGGE gels including ethidium bromide and other nucleic acid stains. This protocol describes the silver staining of gels [10]. An advantage of this technique is that the DGGE bands can later be excised for sequencing without exposure to UV light which may cause DNA strand breaks.

1. Remove the gel from the tank and remove the clamps from the gel sandwich.
2. Place the gel in a staining dish filled with distilled water and remove spacers and glass plates (*see* **Note 10**).
3. Wash twice in distilled water (*see* **Note 11**).
4. Wash in 10 % ethanol for 5 min.
5. Wash in 1 % nitric acid for 3 min.
6. Stain in 0.2 % silver nitrate solution for 20 min (*see* **Note 12**).
7. Wash the gel twice with distilled water.
8. Develop the gel in developer solution until bands appear. Bands should appear after a few minutes.
9. Stop the reaction in 3 % acetic acid.
10. Wash in 10 % ethanol.
11. Take a picture of the gel with an image capture system.

3.4 Sequencing DGGE Bands

1. Excise each DGGE band that is of interest with a clean scalpel.
2. Homogenize each band in 20 µl PCR-grade water with a sterile pipette tip.
3. Amplify the band sequence using PCR with the DGGE PCR primers and 2 µl of the DGGE band solution as template (*refer to* Subheading 3.1).
4. The PCR product can be cloned prior to sequencing.

3.5 Image Analysis

A range of software is available for analyzing DGGE gel images such as Quantity One (Bio-Rad), Image J (NIH, Bethesda, MD, USA), and GelcomparII (Applied Maths, Sint-Martens-Latem, Belgium). These programs use band-finding algorithms to detect and statistically analyze bands on a DGGE gel (Fig. 3). Detection of a band will depend on its intensity which can be influenced by the amount of DNA loaded onto the gel. Therefore, normalization of DGGE profiles either relative to the total intensity of the lane or to a control sample is required. The next step involves the binning of bands with the same mobility and generating a binary matrix based on the presence (1) and absence (0) of bands. From this, each DGGE profile can be compared in a pair-wise manner where similarity coefficients such as the Dice or Jaccard coefficients are

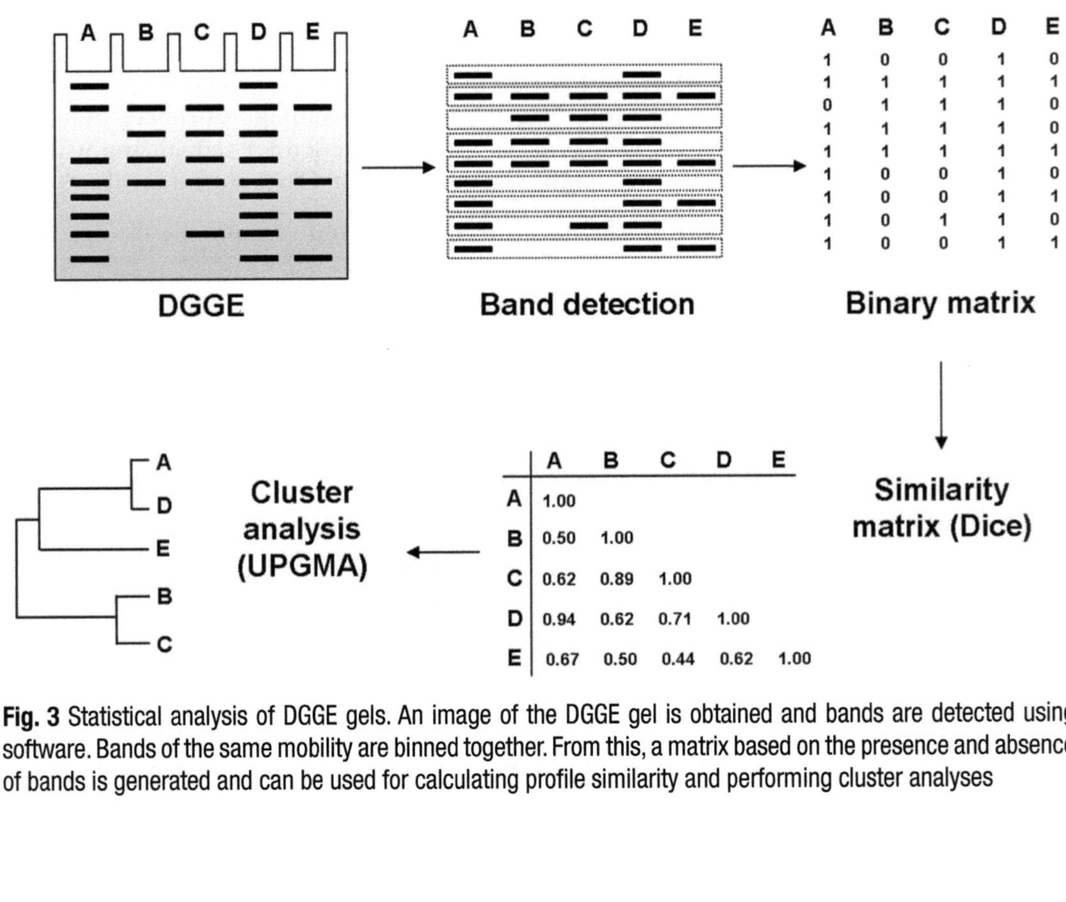

Fig. 3 Statistical analysis of DGGE gels. An image of the DGGE gel is obtained and bands are detected using software. Bands of the same mobility are binned together. From this, a matrix based on the presence and absence of bands is generated and can be used for calculating profile similarity and performing cluster analyses

used to calculate a similarity matrix. The results of this analysis can then be used to generate dendrograms with the help of clustering algorithms such as unweighted pair-wise grouping with mathematical averages (UPGMA).

Measurements of band intensity can also be used to calculate coefficients such as the Simpson and Shannon indices. These indices take into account both species richness and relative abundance. However, it has been previously suggested that DGGE band intensity should not be interpreted as a quantitative measure of the relative abundance of a species [11]. This is because band intensity is also determined by PCR bias, co-migration of multiple species as well as variation in 16S rRNA gene copy number.

One of the limitations in the statistical analysis of DGGE gels is that very large datasets cannot be easily compared. Most commercially available DGGE systems only allow for approximately 20–24 samples to be run per gel. Slight differences in the denaturing gradients between gels will lead to differences in the migration of bands. This means that inter-gel comparisons are not as reliable as comparisons between samples run on the same gel. The use of internal standards may assist in performing comparisons across multiple gels [12, 13].

4 Notes

1. A number of protocols and commercial kits are available to extract DNA from environmental samples such as the FastPrep Spin Kit for Soil (Bio101, La Jolla, CA, USA).
2. Primers F-968-GC (CGC CCG GGG CGC GCC CCG GGC GGG CGG GGG CAC GGG GGG AAC GCG AAG AAC CTT AC) and 1401r (CGG TGT GTA CAA GAC CC) target the V6-V8 segment of the 16S rRNA gene [14]. Other 16S rRNA primer sets are available that are compatible with DGGE such as 341F/534R [1]. Changing primers may require also changing the concentration of the denaturing gradient in order to get an optimal separation of DNA fragments on the gel.
3. Cleaning the glass plates prior to casting increases the quality of the gel and prevents air bubbles from forming during casting.
4. In order to stop the gel leaking after it is cast, it is important to ensure that the glass plates are flush. While the casting sponges will create a seal along the base of the glass plates, 1 % agarose in 1× TAE buffer can be applied to the outside of the plates to provide an additional seal.
5. Mild heating of the 100 % denaturation stock solution prior to adding the acrylamide will help dissolve the urea.

6. Prepare fresh stocks of the denaturant solutions (Table 2) and APS as this will improve the quality of the gel.
7. When the glass plates are added to the sandwich core, it is important to test the seal by adding 1× TAE to the upper buffer compartment. If it leaks then remove the plates and reattach them.
8. At this point it may be helpful to check that the circuit is working correctly by attaching the electrical leads and turning on the current. Turn current off and allow the electrophoresis tank to reach 60 °C.
9. The amount of DNA loaded depends on the expected complexity of the profile. For complex environmental samples load between 200 and 500 ng of DNA. Less DNA needs to be loaded for simple microbial communities or where single bands are expected such as when running DNA from 16S rRNA clones.
10. Touching the gel will cause defects in staining therefore gloves must be worn and the gel should not be touched except at the "fringe" of the gel (where the wells are) if necessary.
11. Transfer of the gel between staining steps can be done by sliding a plastic transparency underneath the gel. This can prevent the gel from breaking. Alternatively commercial products such as Gelbond PAG film (FMC Bioproducts, Rockland, ME, USA) can be used.
12. Silver nitrate is light sensitive. The staining solution as well as the gel should be protected from light during the silver nitrate staining step and after staining is completed.

References

1. Muyzer G, de Waal EC, Uitterlinden AG (1993) Profiling of complex microbial populations by denaturing gradient gel electrophoresis analysis of polymerase chain reaction-amplified genes coding for 16S rRNA. Appl Environ Microbiol 59: 695–700
2. Nielsen DS et al (2003) Case study of the distribution of mucosa-associated Bifidobacterium species, Lactobacillus species, and other lactic acid bacteria in the human colon. Appl Environ Microbiol 69:7545–7548
3. Thompson CL, Holmes AJ (2009) A window of environmental dependence is evident in multiple phylogenetically distinct subgroups in the faecal community of piglets. FEMS Microbiol Lett 290:91–97
4. Shen J et al (2006) Molecular profiling of the Clostridium leptum subgroup in human fecal microflora by PCR-denaturing gradient gel electrophoresis and clone library analysis. Appl Environ Microbiol 72:5232–5238
5. Walter J et al (2001) Detection of Lactobacillus, Pediococcus, Leuconostoc, and Weissella species in human feces by using group-specific PCR primers and denaturing gradient gel electrophoresis. Appl Environ Microbiol 67:2578–2585
6. Favier CF et al (2002) Molecular monitoring of succession of bacterial communities in human neonates. Appl Environ Microbiol 68: 219–226
7. Thompson CL, Wang B, Holmes AJ (2008) The immediate environment during postnatal development has long term impact on gut community structure in pigs. ISME J 2: 739–748
8. Thompson CL et al (2010) Community dynamics in the mouse gut microbiota: a possible role for IRF9-regulated genes in community homeostasis. PLoS One 5:e10335

9. Smalla K et al (2001) Bulk and rhizosphere soil bacterial communities studied by denaturing gradient gel electrophoresis: plant-dependent enrichment and seasonal shifts revealed. Appl Environ Microbiol 67: 4742–4751
10. Sambrook JR, Russell DW (2001) Molecular cloning: a laboratory manual. Cold Spring Harbor Press, New York
11. Murray AE, Hollibaugh JT, Orrego C (1996) Phylogenetic compositions of bacterioplankton from two California estuaries compared by denaturing gradient gel electrophoresis of 16S rDNA fragments. Appl Environ Microbiol 62:2676–2780
12. Neufeld JD, Mohn WW (2005) Fluorophore-labeled primers improve the sensitivity, versatility, and normalization of denaturing gradient gel electrophoresis. Appl Environ Microbiol 71:4893–4896
13. Tourlomousis P et al (2009) PCR-denaturing gradient gel electrophoresis of complex microbial communities: a two-step approach to address the effect of gel-to-gel variation and allow valid comparisons across a large dataset. Microb Ecol 59:776–786
14. Nubel U et al (1996) Sequence heterogeneities of genes encoding 16S rRNAs in Paenibacillus polymyxa detected by temperature gradient gel electrophoresis. J Bacteriol 178:5636–5643

Chapter 5

Terminal Restriction Fragment Length Polymorphism (T-RFLP) Profiling of Bacterial 16S rRNA Genes

Catherine A. Osborne

Abstract

T-RFLP profiling is a very effective method for comparing many samples in an environmental microbiology study, because fingerprints of microbial diversity can be generated in a sensitive, reproducible, and cost-effective manner. This protocol describes the steps required to generate T-RFLP profiles of the dominant members of a bacterial community, by PCR amplification of the bacterial 16S rRNA genes and three restriction endonuclease digests to generate three different profiles for each sample. The generation of multiple profiles per sample provides enough information to confidently differentiate rich environmental bacterial communities.

Key words T-RFLP profiles, Microbial community profiling, Bacterial 16S rRNA gene PCR, Restriction endonucleases

1 Introduction

Terminal Restriction Fragment Length Polymorphism (T-RFLP) was originally developed as a technique to rapidly generate a profile, or snapshot, of the microbial diversity present in environmental samples [1–4]. It is more sensitive and reproducible than a number of other microbial community profiling methods because an internal size standard is run with every sample, so samples from separate runs are still comparable [5–9]. The generation of reproducible and cost-effective profiles, or fingerprints, of the major members of microbial communities makes the T-RFLP profiling technique very effective for comparing many samples in an environmental microbiology study, especially when resources are limited [10].

This protocol describes the steps required to generate T-RFLP profiles of the major bacterial community members, starting at the amplification of the bacterial 16S rRNA genes present in DNA extracted from the samples of interest. The Polymerase Chain Reaction (PCR) amplification requires one oligonucleotide primer

Ian T. Paulsen and Andrew J. Holmes (eds.), *Environmental Microbiology: Methods and Protocols*, Methods in Molecular Biology, vol. 1096, DOI 10.1007/978-1-62703-712-9_5, © Springer Science+Business Media, LLC 2014

to have an attached fluorophore, which is incorporated into all of the amplification products, or amplicons. The amplicons are then digested with a restriction endonuclease (RE), which cuts all of the amplicons at a specific recognition sequence of four base pairs. REs should be chosen to differentiate the range of taxa present in the samples. For samples of relatively low richness, such as gut microbial communities, some sequencing information and one of the available online tools (i.e., TRiFle or DRAT) can help to choose an appropriately discriminating RE [11, 12]. For richer samples, such as soil, where most of the peaks in the profile are likely to correspond to multiple taxa [5, 13], the generation of many profiles per sample, each with a different RE with a unique recognition cut site, provides enough information to actually differentiate bacterial communities [14, 15]. The fragmented amplicons are separated by capillary electrophoresis, and the terminal fragments with the incorporated fluorophore, termed the terminal restriction fragments (T-RFs), are detected and differentiated according to their length, in nucleotides.

The protocol presented here uses three different REs to generate three different bacterial 16S rRNA gene T-RFLP profiles for each sample (Fig. 1). After the generation and analysis of the

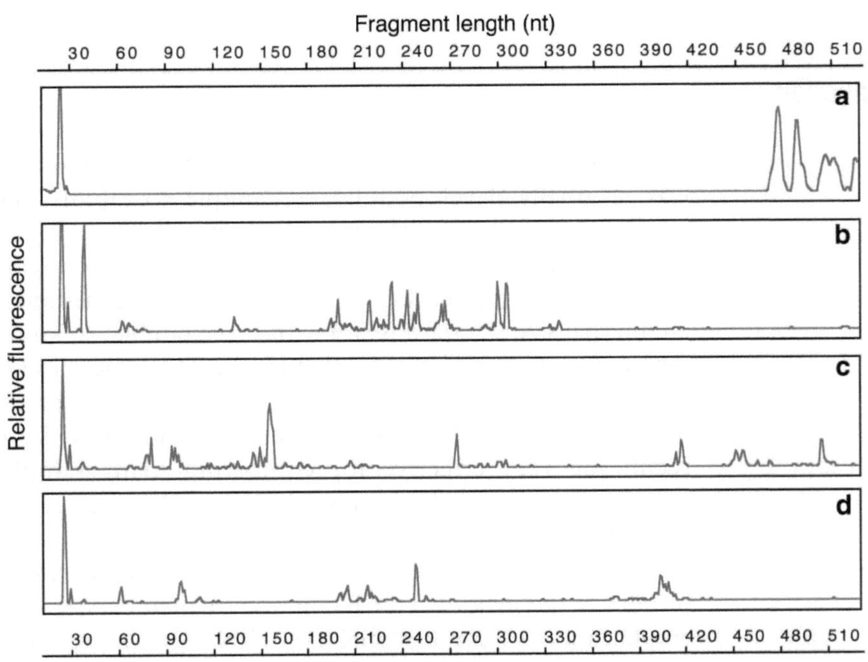

Fig. 1 T-RFLP profiles generated from DNA extracted from a soil sample with the primers FAM27f and 519r, (**a**) without a digestion step where the undigested PCR products are visible (>470 nt), and (**b–d**) after digestion with one of three different restriction endonucleases, showing how the same pool of amplicons gives different profiles because each RE cuts at different recognition sequences (Table 2)

T-RFLP profiles from a statistically powerful dataset, if there are T-RFs of interest that need to be identified, a small amount of DNA sequence information may be able to putatively identify the bacteria responsible for these T-RFs, or peaks [15]. It is preferable to generate sequence data from the same samples as the T-RFLP profiles, rather than assigning identity based on information from online databases, as it proves the occurrence of those 16S rRNA genes in the samples. This approach should be used with caution, however, as more than three REs may be required to confidently assign a taxonomic identity to T-RFs in profiles of rich communities, because many taxa will share common restriction cut sites [16].

The following method can easily be adapted for use on RNA extracts [17], and with the substitution of specific oligonucleotide primers, genes encoding different functions [8] or other phylogenetic groups [18] can be targeted and profiled. A further extension is the multiplexing of different colored fluorophores on various primer sets to assess multiple phylogenetic groups, such as bacteria, archaea, and fungi, within a single profile [19, 20].

2 Materials

Prepare all solutions using ultrapure water (either double-distilled or MilliQ or Barnsted purification systems). All pipette tips and plastic tubes should be nuclease-free and DNA-free. Filter pipette tips are used throughout.

2.1 Water and Buffers

1. Filter-sterilized, UV-cross-linked water: Autoclaved, ultrapure water should be filter-sterilized, into 10 or 15 mL polypropylene tubes (*see* **Note 1**), using aseptic technique and a sterile syringe fitted with a sterile 0.2 μm filter. The tubes should be capped tightly and subjected to a UV light for 15–30 min (*see* **Note 2**).
2. 1 M Tris buffer, pH 8.0: Add 48.4 g Tris [MW 121.14] to a glass beaker on a magnetic stirring plate containing approximately 250 mL of ultrapure water, whilst stirring. When the Tris has dissolved fully, adjust the pH of the solution to 8.0 with concentrated HCl. Transfer the liquid to a clean, glass measuring cylinder and add ultrapure water up to 400 mL. Transfer to a 500 mL reagent bottle (e.g., Schott), and sterilize by autoclaving, with the lid slightly loose, for 20 min at 121 °C.
3. 10 mM Tris buffer: Make 10 mM Tris by diluting 1 M Tris buffer, pH 8.0, 1:100 in autoclaved, ultrapure water. Filter-sterilize 10 mM Tris into polypropylene tubes and UV-cross-link (*see* **Note 3**).

Table 1
DNA sequences of the oligonucleotide primers required for this protocol

Primer name	DNA sequence (5′→3′)	Purity grade
FAM27f	[FAM] GAG TTT GAT CMT GGC TCA G	HPLC
BAC519r	GWA TTA CCG CGG CKG CTG	Transfection

The purity grades (according to Geneworks, an oligonucleotide supplier) are: Transfection grade, which is purified to a higher level than the standard desalted grade to remove trace amounts of organics and salts and short failure sequences; and HPLC purified, which is purified by size using Reverse-Phase High Performance Liquid Chromatography (HPLC), so is free from all organics and salts and contain 97–99 % of the full length product. The fluorescent dye, 6-carboxyfluorescein (FAM), is attached at the 5′ end of the FAM27f primer

2.2 PCR Amplification, Including Incorporation of the Fluorescent Label

1. Oligonucleotide primers: FAM27f, labelled at the 5′ terminus with the fluorescent dye, 6-carboxyfluorescein (FAM; *see* **Note 4**), and BAC519r (*see* **Note 5**; Table 1). Manufactured primers are supplied in a lyophilized form and should be pelleted before resuspending (approximately 15–30 s at maximum speed in a benchtop centrifuge). To make a 100 μM stock solution of the lyophilized primers, add [10× nmol] μL of 10 mM Tris buffer. After vortexing (15 s) and spinning again, allow the pellets to rehydrate on the bench for 30–60 min. A 10 μM working solution can be made by diluting the 100 μM stock solution 1:10 with 10 mM Tris buffer in a 1.5 mL plastic tube. Store both the 100 μM stock and 10 μM working solutions at −20 °C, protected from light (*see* **Note 4**).
2. DNA Polymerase: Available from many molecular biology suppliers, often supplied with an optimal buffer as a 10× concentrate. This protocol uses Qiagen HotStarTaq DNA Polymerase (*see* **Note 6**). Store polymerase and buffer at −20 °C.
3. Deoxynucleotide Triphosphates (dNTPs): Also available from many molecular biology suppliers. Stock solutions are often supplied as the four dNTPs separately (dATP, dCTP, dGTP, and dTTP), which are then mixed together to create a dNTP working solution. The recipe in this procedure requires a working solution that contains 25 mM of each dNTP. If the purchased stock solutions are 100 mM, then add equal amounts of each of the four solutions to a 1.5 mL plastic tube to produce the required concentration of each dNTP. If further dilution is required, use the filter-sterilized, UV-treated water. Store the stock and working solutions at −20 °C.

4. DNA extracted from the samples of interest: Use either a commercially available kit (e.g., MoBio PowerSoil DNA Isolation Kit) or a published or lab-developed protocol [21] to extract good quality DNA from the samples of interest. Quantify the concentration of the DNA in the extracts (*see* **Note 7**) and dilute each of the extracts to a working concentration of 1.0 ng/μL using 10 mM Tris buffer (*see* **Note 8**). Store the 1.0 ng/μL working solution and the stock DNA extracts at −20 °C, or −80 °C if available.
5. Thermal cycler: a programmable peltier device that allows PCR amplification by rapidly warming and cooling of 24 or 96 tube-positions. Available from a number of companies, including Bio-Rad and Hybaid.

2.3 Restriction Digestion of the PCR Amplicons and Separation of the Fragments

1. Restriction endonucleases (REs): HinfI, MspI, and Sau96I. This protocol uses the enzyme preparations available from New England Biolabs (NEB), which are each supplied with an optimal buffer as a 10× concentrate (Table 2). Store at −20 °C and thaw on ice when needed.
2. 75 % (v/v) Isopropanol: tip 75 % of the water out of a tube of filter-sterilized, UV-cross-linked water and refill with molecular biology-grade isopropanol (also known as isopropyl alcohol or propan-2-ol) and mix by inversion.
3. Loading buffer: 100:1 HiDi Formamide/LIZ600 size standard (both from Applied Biosystems). Make up enough of the 100:1 solution for the number of samples plus 10 % and mix by vortexing.
4. ABI 3130xl Genetic Analyzer (Applied Biosystems): a capillary electrophoresis instrument for separation of DNA fragments, based on their size and fluorescence (*see* **Note 9**).

Table 2
Three restriction endonucleases used for digesting FAM-labelled PCR amplicons

Restriction endonuclease	Recognition cut site	Stock conc. (NEB)	Buffer no. (NEB)	Active temp.	Inactivation
HinfI	GANTC	10 U/μL	2	37 °C	80 °C, 20 min
MspI	CCGG	20 U/μL	2	37 °C	65 °C, 20 min
Sau96I	GGNCC	5 U/μL	4	37 °C	80 °C, 20 min

The stock concentrations (conc.) are for the restriction endonucleases obtained from New England Biolabs (NEB) and the buffer number (no.) is the optimal NEB designated buffer that comes with the purchased enzyme

3 Methods

3.1 PCR Amplification, Including Incorporation of the Fluorescent Label

1. In a 1.5 or 2 mL tube, prepare a PCR master mix based on the recipe given in Table 3 (*see* **Note 10**), allowing for at least three replicate reactions for each sample, positive and negative control reactions, plus an extra 10 % (*see* **Note 11**).
2. Aliquot 1 μL of the samples (1 ng/μL) and the controls into 0.2 mL plastic tubes (*see* **Note 12**), and then aliquot 49 μL of the PCR master mix into each tube.
3. Briefly centrifuge the tubes (15 s) so the DNA and the reagents are at the bottom, with no air bubbles.
4. Place the 0.2 mL tubes in the thermal cycler, and program it to run at 95 °C for 15 min to activate the HotStarTaq DNA Polymerase, followed by 25 cycles of 94 °C for 30 s, 48 °C for 30 s, and 72 °C for 60 s, with a final extension step of 72 °C for 10 min (*see* **Note 13**).
5. Check all of the PCR products by gel electrophoresis (*see* **Note 14**). Samples and positive controls should have only a single product of the correct size (i.e., 500 bp) while negative controls should not contain any PCR products (*see* **Note 15**).
6. The triplicate PCR reactions for one sample (3 × 50 μL) should be pooled into a single tube and then purified using a commercial PCR cleanup kit, e.g., QIAquick PCR Purification Kit

Table 3
Recipes for the PCR master mix, to amplify and incorporate the terminal FAM label in 1, 10, or 25 PCR reactions, if using Qiagen HotStarTaq DNA Polymerase

PCR reagent	Starting conc.	Final conc. in PCR	1 PCR (49 μL)	10 PCRs (490 μL)	25 PCRs (1,225 μL)
H_2O (filtered and UV-treated)			41.0 μL	410 μL	1,025 μL
10× Buffer	15 mM Mg^{2+}	1.5 mM Mg^{2+}	5.0 μL	50 μL	125 μL
FAM27f	10 μM	0.2 μM	1.0 μL	10 μL	25 μL
BAC519r	10 μM	0.2 μM	1.0 μL	10 μL	25 μL
dNTPs	25 mM each	250 μM each	0.5 μL	5 μL	12.5 μL
HotStarTaq	5 U/μL	0.05 U/μL	0.5 μL	5 μL	12.5 μL

The starting and final concentrations of each reagent are given, as are the volumes required for 1, 10, or 25 PCRs. The 10× buffer is the optimal Qiagen HotStarTaq buffer that comes with the purchased HotStarTaq. The concentration of Mg^{2+} in the final PCR can influence the amplification, so the final concentration of this key component in the buffer is shown, and no additional Mg^{2+} is necessary

Table 4
Recipes for the RE master mixes for the three REs: (a) HinfI, (b) MspI, and (c) Sau96I, if obtained from New England Biolabs

Reagent	1 Digest	10 Digests	25 Digests
(a)			
H_2O	2.0 μL	20 μL	50 μL
10× Buffer no. 2	2.5 μL	25 μL	62.5 μL
HinfI [10 U/μL]	0.5 μL	5 μL	12.5 μL
(b)			
H_2O	2.25 μL	22.5 μL	56.25 μL
10× Buffer no. 2	2.50 μL	25 μL	62.5 μL
MspI [20 U/μL]	0.25 μL	2.5 μL	6.25 μL
(c)			
H_2O	1.5 μL	15 μL	37.5 μL
10× Buffer no. 4	2.5 μL	25 μL	62.5 μL
Sau96I [5 U/μL]	1.0 μL	10 μL	25 μL

5 μL of each master mix, containing 5 U of RE, is added to 20 μL of solution containing 20 ng of FAM-labelled amplicons

(Qiagen) or Wizard SV PCR Purification Kit (Promega), according to the manufacturer's instructions (*see* **Note 16**).

7. Quantify each of the purified PCR products (*see* **Note 7**), before proceeding to the restriction digestion (*see* **Note 17**).

3.2 Restriction Digestion of the PCR Amplicons and Separation of the Fragments

1. In three 1.5 mL tubes on ice, prepare 3 RE master mixes, one for each RE, based on the recipes given in Table 4 (*see* **Note 10**). Prepare enough for the total number of samples and controls, plus 10 % extra, and keep on ice.
2. In a 96-well semi-skirted plate on ice (*see* **Note 18**), add 20 ng of each PCR product to 3 wells. Make the volume up to 20 μL in each well with filter-sterilized, UV-treated water.
3. Add 5 μL of each of the 3 RE master mixes to a single well for each sample (*see* **Note 19**).
4. Seal the plate and gently bang it on the bench, to mix the amplicons and the RE at the bottom of the wells.
5. Place the plate in the thermal cycler, and program it to run at 37 °C for 4 h, followed by 80 °C for 20 min to inactivate the REs. After removing the plate, set the thermal cycler to 70 °C.
6. Add 100 μL 75 % isopropanol to each digest, vortex the plate, and allow it to sit on the bench for 30 min, protected from light.
7. Centrifuge the plate for 30 min at maximum speed, in a centrifuge with the ability to spin 96-well plates (*see* **Note 18**). Remove the plate seal and discard the supernatant.

8. Place the plate outside down on paper towel and place back into the centrifuge. Pulse for 15 s, so the residual supernatant is spun out of the wells and onto the paper towel.
9. Place the plate in the thermal cycler, with the lid open, for 5 min at 70 °C, to evaporate the residual isopropanol (*see* **Note 20**).
10. Remove the plate from the thermal cycler and add 10 μL of 100:1 HiDi formamide/LIZ600 size standard to each well. After removing the plate, set the thermal cycler to 95 °C.
11. Vortex the plate (15 s) and centrifuge again (15 s).
12. Place the plate in the thermal cycler at 95 °C for 3 min, to denature the fragmented amplicons.
13. Immediately after the denaturation step, place the plate on ice before transferring into the Genetic Analyzer cartridges, which consist of a black tray, grey insert, and white lid clips, and then into the ABI 3130xl Genetic Analyzer (*see* **Note 21**).

3.3 Data Analysis

1. Raw T-RFLP profile data can be exported from the ABI 3130xl Genetic Analyzer and analyzed with GeneMapper software (Applied Biosystems), using the AFLP settings. Profiles generated with different REs are kept in separate GeneMapper projects because they are not comparable.
2. Within GeneMapper, each user should determine what is appropriate for their dataset, but a good start is setting the bin width to 1.5 nt and trimming the profiles to between 50 nt and 550 nt, based on the presence of shorter peaks in negative control profiles and the absence of fragments longer than 550 nt from this primer pair.
3. Files of the peak area and peak size (length in nt) are exported as comma-separated .csv files (*see* **Note 22**) and then can be imported into a number of different programs for further analysis (*see* **Note 23**).
4. For most microbial communities, if pairwise comparisons of the profiles are required then the application of a threshold is recommended before profiles are compared, so that T-RFs that may only be present in some samples due to loading of more fluorescent product are removed (*see* **Note 24**).
5. For rich microbial communities, such as soil, it is also preferable to use presence/absence scoring of common T-RFs in T-RFLP profiles, such as the Sorensen's pairwise distance matrix, rather than relative abundance calculations, because the relative abundance may be influenced by unquantifiable biases in the preparation steps of DNA extractions and PCR amplifications (*see* **Note 25**).

4 Notes

1. Tubes made of polypropylene are preferable to those made of polystyrene because UV light penetrates polypropylene better than polystyrene.
2. Take care to prevent skin and eye exposure to UV light. Designated UV-cross-linking instruments, such as a Spectrolinker XL-1000 (Spectronics Corporation), set at 254 nm and 1,200 mJ/cm^2, can be used, but the UV light in a biosafety cabinet or a gel documentation station can also be used to cross-link any contaminating double-stranded DNA (dsDNA) present.
3. This protocol resuspends the primers and dilutes the DNA in a 10 mM Tris buffer, rather than a TE (10 mM Tris, 1 mM EDTA) buffer, to prevent the addition of EDTA to the PCR, which can chelate the Mg^{2+} present, thus affecting the optimum Mg^{2+} concentration for DNA Polymerase action.
4. Because the FAM fluorophore is photosensitive, the stock and working solutions of the FAM-labelled primer, and any subsequent FAM-incorporated PCR products (amplicons), should be kept away from direct light as much as practically possible, i.e., by wrapping or covering with foil and/or using amber-colored 1.5 mL plastic tubes to store primer solutions or combine PCR master mixes.
5. Other primers can be used but this primer pair is preferable for bacterial 16S rRNA gene amplification for three reasons: it is predicted to bind universally to most known bacterial 16S rRNA genes; if there are significant amounts of undigested PCR product then it can be detected on the profiles as large peaks at approximately 470–530 nt (Fig. 1a); and it avoids PCR-generated artifacts that can occur with 1492r, another commonly used reverse primer for bacterial 16S rRNA gene amplification [22].
6. This protocol uses HotStarTaq DNA Polymerase (Qiagen) because it is a highly purified DNA Polymerase preparation that does not generate positive bacterial 16S rRNA gene PCR products in the negative controls, which can occur in preparations of lower quality, where genomic DNA from the expression host has not been fully removed [23, 24]. The "hot start" requirement (15 min at 95 °C) also prevents the amplification of spurious, nonspecific PCR products.
7. Quantify DNA with the Quant-iT PicoGreen dsDNA Quantitation reagent (Invitrogen), according to the manufacturer's instructions, or by measuring the solution's absorbance at 260 nm with a NanoDrop instrument (Thermo Scientific).

8. DNA replication by DNA Polymerase in a PCR can be inhibited if a high concentration of enzymatic inhibitors remains in the DNA extract. Diluting the DNA extract to a concentration of 1.0 ng/μL should overcome this potential obstacle and provide equivalent amount of template in all of the PCRs.
9. As most capillary electrophoresis instruments are purchased and used for Sanger sequencing, many commercial sequencing facilities are able to run T-RFLP profiles as well, often under a service called "Fragment Analysis." If the service provider has a Genetic Analyzer, they should also have a license for the GeneMapper software (also from Applied Biosystems) which will allow them to supply you with the results, in the form of electropherograms, or profiles (e.g., Fig. 1), as well as a spreadsheet output with the length (in nucleotides) and fluorescence area of the aligned peaks/T-RFs.
10. Make sure that all reagents are completely thawed and well-mixed (vortex for 15 s and then centrifuge for 5 s) before aliquoting. If using reagents from alternative suppliers, make note of different concentrations of enzyme and the supplied buffer and adjust the volumes added to the master mix accordingly, including potentially adjusting the volume of filter-sterilized, UV-cross-linked water required to get the master mix to the correct final volume.
11. The positive control should consist of DNA extracted from a known bacterial culture, or an environmental sample that has previously been known to amplify with these primers. Negative controls include no addition to the aliquot of PCR master mix and the buffers used to resuspend and dilute the DNA extracts or a blank DNA extraction.
12. 0.2 mL tubes can be purchased as single tubes, or connected as 8-strips or in a 96-well format, any of which can be used in the PCRs, depending on the thermal cycler setup and the number of reactions.
13. Cycle the PCR only 25 times to minimize potential bias in the resulting amplicon pools.
14. Traditional agarose gel electrophoretic equipment, or an E-Gel Electrophoresis System (Invitrogen), should be used to separate the PCR products and a UV transilluminator is required to visualize the PCR products. If traditional gel electrophoresis equipment is available, then loading 5 μL of each PCR and a DNA size standard, e.g., O'GeneRuler 100 bp DNA Ladder (Fermentas), into the wells of a 2 % (w/v) agarose gel and running for 25 min at 120 V will be appropriate for visualizing the PCR products (expected size of approximately 500 bp). Sterile electrophoresis buffer, either TAE buffer (40 mM Tris, pH 8.0 with acetic acid, 1 mM EDTA) or SB buffer (10 mM sodium hydroxide, pH 8.5 with boric acid), should be used to make

the agarose gel. The DNA stain, either SYBR Safe (Invitrogen) or ethidium bromide (final concentration 0.1 μg/μL), can be added to the melted agarose solution before it is poured into the gel mold.

15. Any smears or large bands near the bottom (approx. 20–50 bp long) of the gel can indicate an inefficient PCR reaction, and potentially even the presence of primer-dimer artifacts that can adversely affect the resulting T-RFLP profiles by obscuring true peaks/T-RFs [22].

16. Commercial spin-column kits for PCR purification are the preferred purification method because they remove unincorporated primers and primer-dimer artifacts, which can potentially affect the T-RFLP profiles [22], more effectively than alcohol precipitation. Commercial PCR clean-up kits that dilute the PCR products in a buffer before binding to a spin-column are also preferable to extracting the PCR products from an agarose gel slice, because these kits only lose approximately 30 % of the original PCR product whereas the gel extraction kits lose approximately 60 % of the PCR product. For the final elution step in any of these kits, we recommend waiting 5 min after adding the elution buffer to the spin column, to allow all of the bound DNA to resuspend before the final spin.

17. "Pseudo-TRFs" can arise when single-stranded DNA in the restriction digest make stable dsDNA structures that the REs can also digest [25]. Pseudo-TRFs are important to remove if trying to assign identity to peaks in a profile, such as in samples with low richness. Pseudo-TRFs can be prevented with the addition of a mung bean nuclease digestion step [25].

18. This protocol uses the 96-well format for the restriction digestion and precipitation, because these plates can be placed straight in the Genetic Analyzer, but any of the 0.2 mL tubes (single tubes or 8-strips) can be used at this point. If a centrifuge able to spin 0.2 mL tubes or 96-well plates is not available and a temperature-controlled water bath is available, then the restriction digests and precipitation can occur in 1.5 mL tubes. Once the pellet is resuspended in 100:1 HiDi Formamide/LIZ600, then the fragmented amplicons can be transferred to the wells of an appropriate 96-well semi-skirted plate for the Genetic Analyzer.

19. Use of a multi-dispensing pipetting aid, e.g., Multipette or Repeater Plus (Eppendorf), can make dispensing the same volume into many wells ergonomically easier. Make sure that the tips used are molecular biology-grade, i.e., Biopur Combitips (Eppendorf), and avoid splashing when dispensing to prevent cross-contamination.

20. Most commercial fragment analysis facilities would accept samples as dry pellets at this point in the protocol (*see* **Note 9**).

21. A tab-delimited .txt file needs to be imported onto the ABI 3130xl Genetic Analyzer with the names of the samples and the instrument protocol to use. Use an instrument protocol with a run time of 20 min to get resolution to 600 nt, but shorter runs of 12 or 15 min can be used if you are only interested in fragments up to 500 nt.
22. Peak areas are preferable to peak heights because the peak area represents the actual amount of fluorescence each peak contributes to the profile's total fluorescence [4], i.e., proportions of the total profile/community can be calculated, but some researchers still use peak height.
23. There is a specific software available online for processing T-RFLP data, i.e., T-REX [26], but spreadsheet manipulations can be carried out in Excel (Microsoft) and specific ecological calculations and comparisons can be done with Primer6 software (Primer-E Ltd) or The R Project (http://www.r-project.org). Within The R Project there is a T-RFLP specific package, TRAMPR [27], but a lot of analyses and comparisons can be conducted using the ecology package, Vegan [28].
24. The optimal variable percentage threshold [14] can be determined for a small dataset relatively easily using an Excel spreadsheet, or for a larger dataset using The R Project (an Excel spreadsheet and R scripts for determining the optimal threshold of T-RFLP datasets are available at http://www.researchgate.net/profile/Catherine_Osborne/blog/).
25. It is probably also unwise to calculate univariate indices of diversity, such as evenness and dominance, because such simplified fingerprints of complex communities certainly underestimate the diversity present [29], though this is debated in the literature [30].

Acknowledgement

The author is supported by the Centre for Water Sensitive Cities in the Department of Civil Engineering at Monash University, Clayton, Victoria, Australia.

References

1. Liu W-T et al (1997) Characterization of microbial diversity by determining terminal restriction fragment length polymorphisms of genes encoding 16S rRNA. Appl Environ Microbiol 63:4516–4522
2. Clement BG et al (1998) Terminal restriction fragment patterns (TRFPs), a rapid, PCR-based method for the comparison of complex bacterial communities. J Microbiol Methods 31:135–142
3. Marsh TL (1999) Terminal restriction length polymorphism (T-RFLP): an emerging method for characterizing diversity among homologous populations of amplification products. Curr Opin Microbiol 2:323–327
4. Kitts CL (2001) Terminal restriction fragment patterns: a tool for comparing microbial communities and assessing community dynamics. Curr Issues Intest Microbiol 2:17–25

5. Moeseneder MM et al (1999) Optimization of terminal-restriction fragment length polymorphism analysis for complex marine bacterioplankton communities and comparison with denaturing gradient gel electrophoresis. Appl Environ Microbiol 65:3518–3525
6. Osborn AM, Moore ERB, Timmis KN (2000) An evaluation of terminal-restriction fragment length polymorphism (T-RFLP) analysis for the study of microbial community structure and dynamics. Environ Microbiol 2:39–50
7. Brodie EL, Edwards S, Clipson N (2002) Bacterial community dynamics across a floristic gradient in a temperate upland grassland ecosystem. Microb Ecol 44:260–270
8. Lueders T, Friedrich MW (2003) Evaluation of PCR amplification bias by terminal restriction fragment length polymorphism analysis of small-subunit rRNA and *mcrA* genes by using defined template mixtures of methanogenic pure cultures and soil DNA extracts. Appl Environ Microbiol 69:320–326
9. Hartmann M et al (2005) Semi-automated genetic analyses of soil microbial communities: comparison of T-RFLP and RISA based on descriptive and discriminative statistical approaches. J Microbiol Methods 61:349–360
10. Orcutt B et al (2009) An interlaboratory comparison of 16S rRNA gene-based terminal restriction fragment length polymorphism and sequencing methods for assessing microbial diversity of seafloor basalts. Environ Microbiol 11:1728–1735
11. Junier P, Junier T, Witzel K-P (2008) TRiFLe, a program for in silico terminal restriction fragment length polymorphism analysis with user-defined sequence sets. Appl Environ Microbiol 74:6452–6456
12. Roberts DM et al (2012) Directed terminal restriction analysis tool (DRAT): an aid to enzyme selection for directed terminal-restriction fragment length polymorphisms. Methods Ecol Evol 3:24–28
13. Rösch C, Bothe H (2005) Improved assessment of denitrifying, N_2-fixing and total community bacteria by terminal restriction fragment length polymorphism analysis using multiple restriction enzymes. Appl Environ Microbiol 71:2026–2035
14. Osborne CA et al (2006) New threshold and confidence estimates for terminal restriction fragment length polymorphism analysis of complex communities. Appl Environ Microbiol 72:1270–1278
15. Osborne CA, Peoples MB, Janssen PH (2010) Detection of a reproducible, single-member shift in soil bacterial communities exposed to low levels of hydrogen. Appl Environ Microbiol 76:1471–1479
16. Dunbar J, Ticknor LO, Kuske CR (2001) Phylogenetic specificity and reproducibility and new method for analysis of terminal restriction fragment profiles of 16S rRNA genes from bacterial communities. Appl Environ Microbiol 67:190–197
17. Carson JK et al (2010) Low pore connectivity increases bacterial diversity in soil. Appl Environ Microbiol 76:3936–3942
18. Bennett LT, Kasel S, Tibbits J (2009) Woodland trees modulate soil resources and conserve fungal diversity in fragmented landscapes. Soil Biol Biochem 41:2162–2169
19. Singh BK et al (2006) Use of multiplex terminal restriction fragment length polymorphism for rapid and simultaneous analysis of different components of the soil microbial community. Appl Environ Microbiol 72:7278–7285
20. Bissett A et al (2011) Long-term land use effects on soil microbial community structure and function. Appl Soil Ecol 51:66–78
21. Griffiths RI et al (2000) Rapid method for coextraction of DNA and RNA from natural environments for analysis of ribosomal DNA- and rRNA-based microbial community composition. Appl Environ Microbiol 66: 5488–5491
22. Osborne CA et al (2005) PCR-generated artefact from 16S rRNA gene-specific primers. FEMS Microbiol Lett 248:183–187
23. Tanner MA et al (1998) Specific ribosomal DNA sequences from diverse environmental settings correlate with experimental contaminants. Appl Environ Microbiol 64:3110–3113
24. Zehr JP et al (2003) Nitrogenase genes in PCR and RT-PCR reagents and implications for studies of diverse functional genes. Biotechniques 35:996–1005
25. Egert M, Friedrich MW (2003) Formation of pseudo-terminal restriction fragments, a PCR-related bias affecting terminal restriction fragment length polymorphism analysis of microbial community structure. Appl Environ Microbiol 69:2555–2562
26. Culman S et al (2009) T-REX: software for the processing and analysis of T-RFLP data. BMC Bioinformatics 10:171
27. Fitzjohn RG, Dickie IA (2007) TRAMPR: an R package for analysis and matching of terminal-restriction fragment length polymorphism (TRFLP) profiles. Mol Ecol Notes 7:583–587
28. Oksanen J et al (2010) vegan: community ecology package. R package version 1.17-3. http://CRAN.R-project.org/package=vegan
29. Dunbar J, Ticknor LO, Kuske CR (2000) Assessment of microbial diversity in four southwestern United States soil by 16S rRNA gene terminal restriction fragment analysis. Appl Environ Microbiol 66:2943–2950
30. Bent SJ et al (2007) Measuring species richness based on microbial community fingerprints: the emperor has no clothes. Appl Environ Microbiol 73:2399–2401

Chapter 6

Profiling the Diversity of Microbial Communities with Single-Strand Conformation Polymorphism (SSCP)

Achim Schmalenberger and Christoph C. Tebbe

Abstract

Genetic fingerprinting techniques for microbial community analysis have evolved over the last decade into standard applications for efficient and fast differentiation of microbial communities based on their diversity. These techniques commonly analyze the diversity of PCR products amplified from extracted environmental DNA usually utilizing primers hybridizing to suspected conserved regions of the targeted genes. In comparison to the more commonly applied terminal restriction fragment length polymorphism (TRFLP) or denaturing gradient gel electrophoresis (DGGE) techniques, the here-described single-strand conformation polymorphism (SSCP) fingerprinting technique features some advantageous key characteristics. (1) Primers for the polymerase chain reaction (PCR) do only need minimal 5′-end alterations; (2) SSCP is adaptable to high throughput applications in automated sequencers; and (3) a second dimension in the SSCP gel electrophoresis can be implemented to obtain high resolution 2D gels. One central key requirement for SSCP gel electrophoresis is a tight temperature control. Gels that run at different temperatures will produce entirely different fingerprints. This can be exploited for an improved analysis of highly diverse communities by running the same template at different temperatures or by 2D-SSCP gel electrophoresis.

Key words PCR-SSCP, ssDNA, DNA conformation, microbial diversity, community profile, DNA fingerprint, 16S rRNA gene

1 Introduction

Genetic fingerprinting techniques provide an important means to display the diversity of microbial community members and thus allow the comparison of their composition from different environmental samples. Currently applied methods in environmental microbiology rely on the analyses of products amplified by the polymerase chain reaction (PCR) from extracted DNA. Typically such amplifications are conducted with PCR primers that hybridize to phylogenetically conserved regions of a target gene and thereby amplify a mixture of PCR products, which require further processing in order to generate genetic fingerprints by electrophoretic separation techniques. Electrophoretic separation can be achieved

Ian T. Paulsen and Andrew J. Holmes (eds.), *Environmental Microbiology: Methods and Protocols*, Methods in Molecular Biology, vol. 1096, DOI 10.1007/978-1-62703-712-9_6, © Springer Science+Business Media, LLC 2014

either under denaturing conditions, if the DNA fragments under investigation have a different size, or under native (non-denaturing) conditions. The native conditions may be combined with increasingly denaturing conditions either by chemical or temperature gradients. Under fully native conditions, it is in fact possible to differentiate between DNA fragments of the same length as the electrophoretic mobility is then affected by the secondary structure of the DNA molecule which is strongly influenced by the nucleotide sequence of the four bases, adenine (A), thymine (T), guanine (G), and cytosine (C).

Among the genetic fingerprinting techniques, the terminal restriction fragment length polymorphism (TRFLP) technique is most commonly applied when DNA fragments are separated by size. Since amplification of community DNA typically generates DNA fragments of a similar length, these products must first be digested with restriction endonucleases to obtain fragments of different lengths. To detect terminal restriction fragments (TRF), one or both PCR primers need a fluorescent dye as a label [1]. Denaturing gradient gel electrophoresis (DGGE [2]), temperature gradient gel electrophoresis (TGGE [3]), and single-strand conformation polymorphism (SSCP [4, 5]) separate whole PCR products despite their lack of size differences. DNA separations in DGGE and TGGE are based on a denaturing or temperature gradient and a GC-clamp on one of the primers in use. Such a clamp which is typically composed of additional 30 nucleotides [2] allows the maintenance of part double-stranded (ds) DNA even when the opposite DNA strands of the PCR-targeted gene are completely separated by denaturation. In contrast, SSCP works with single-stranded (ss) DNA. PCR products are denatured by heat into ssDNA prior to electrophoresis in non-denaturing gels. The impact of the base sequence on the electrophoretic mobility can in fact be very strong, and substitutions of a single base in a 300-base-long sequence are detectable [4]. In contrast to DGGE and TGGE, fingerprinting with SSCP only requires minimal modifications of the primers in order to generate antisense-free ssDNA.

Originally, SSCP was developed to detect gene polymorphism in human DNA [4] and mutations by comparing PCR products [6]. In these pioneering works, dsDNA was used for the gel electrophoresis. In order to obtain conformation-specific information, dsDNA was denatured with formamide and heat to achieve strand separation and thus obtain ssDNA. However, during electrophoresis, reannealing occurred among the complementary ssDNA with the result that typically for each DNA fragment analyzed, three bands occurred: two generated by the complementary ssDNA molecules and one caused by the reannealed dsDNA. The method was also applied to characterize microbial community by PCR, but due to the heterogeneity of the amplified PCR products from environmental DNA, the patterns of even simply structured communities were very complex. While the different mobility of opposite ssDNA strands is doubling the number of detectable signals, the formation of dsDNA molecule heteroduplexes built by reannealing of the complementary or almost complementary ssDNA [7] introduces a multitude of additional signals in the community profile. In microbial ecology this classical approach has its merits to screen differences between pure culture isolates or cloned 16S rRNA gene libraries, as demonstrated by Tebbe et al. [8]. However, for the analyses of diverse microbial communities, the extensive level of heteroduplex formations makes the method unfeasible. Ideally, each community member should only generate one signal (band or peak, depending on the detection) and not an undefined amount of products and side products [5]. The selective removal of one of the DNA strands from the dsDNA PCR product was the solution of the problem, as published by Schwieger and Tebbe in 1998 [5]. This was achieved by the introduction of a 5′-end phosphorylation of one of the two PCR primers that allows the digestion of the phosphorylated DNA strand with lambda exonuclease prior to the SSCP gel electrophoresis. The feasibility of this modified approach for microbial community analysis was first demonstrated for analyzing rhizosphere soils and composts or the gut contents of invertebrates [5, 9]. Compared to the more commonly applied fingerprinting techniques, i.e., TRFLP and DGGE, the introduction of the 5′-end phosphorylation has some significant advantages because it is much smaller than a fluorescent dye (TRFLP) and the abovementioned GC-clamp (DGGE).

Since its introduction to environmental microbiology, several distinct applications have been developed for the SSCP technique. Nested PCR applications have widened the spectrum of phylogenetic groups that can be displayed via SSCP and increase the sensitivity of detection [10] in direct comparison to the profiling of the dominant members of the bacterial community. The adaptation to automated sequencing machines as capillary electrophoresis (CE) SSCP has established a high throughput application comparable to TRFLP [11, 12]. In addition to fingerprinting, DNA sequencing of the silver-stained bands from SSCP gels became feasible, allowing to characterize the contributors to the microbial community by phylogenetic analyses of the targeted genes [5, 13]. The transfer of single-stranded DNA from the acrylamide gels onto nylon membranes is another option to identify specific DNA sequences from SSCP profiles by means of Southern hybridization using gene probes [14, 15]. And more recently, the SSCP technique was extended to create two-dimensional gels by choosing different temperature settings for each dimension creating a separation based on potentially two different conformations for each molecule [16].

In the following section we describe the SSCP method for bacterial community analyses, as it is has evolved in different laboratories during 12 years of its use in environmental microbiology.

2 Materials

All solutions should be made up with ultrapure (18 MΩ) or bi-distilled water with the exception of the 1× electrophoresis buffer and the gel staining solutions where deionized water is sufficient.

2.1 PCR

1. Thermocycler for PCR with wells for 0.2 mL micro reaction (PCR) tubes.
2. 0.2 mL PCR tubes, DNA-free (pre-sterilized or autoclaved).
3. Clean micropipettes and DNA-free barrier tips.
4. PCR tube rack on ice.
5. HotMaster *Taq* polymerase (5 PRIME) with supplied buffer (*see* **Note 1**).
6. $MgCl_2$ solution (5 PRIME).
7. DNA-free water (*see* **Note 2**).
8. 0.5 μM primers; standard universal 16S, COM1 (5′CAGCAGCCGCGGTAATAC3′) and COM2-PH (5′CC GTCAATTCCTTTGAGTTT3′ with a phosphate group at the 5′-end) (*see* **Note 3**) supplied by Invitrogen, Metabion, and other companies.
9. Deoxynucleotides 200 μM each.
10. Optional: enhancer solution (5 PRIME and other companies) or self-made (1 M betaine, final conc.).
11. Approx. 1 ng template DNA or 0.1–1.0 μL of template DNA (*see* **Note 4**).

2.2 PCR Product Processing

1. Lambda exonuclease (New England Biolabs or GE Life Sciences).
2. PCR purification kit with a small elution volume of 10–12 μL (e.g., Invitek, Fermentas, Qiagen).

2.3 Gel Electrophoresis

1. Electrophoresis power supply with the capacity of at least 500 V (e.g., Biorad Universal Power Supply).
2. LKB 2010 Macrophor system (LKB, Amersham, discontinued) (*see* **Note 5**).
3. 20 cm notched glass plates for Macrophor (Amersham, now GE Life Sciences).
4. 0.4 mm spacer strips, 40 cm.
5. 0.4 mm comb.
6. Eight large bulldog clamps.
7. Horizontal gel pouring stand with levelling feet (Amersham, discontinued) or level laboratory bench.

8. Vacuum chamber or vacuum flask with vacuum pump (e.g., water pump).
9. MultiTemp II or III recirculating chiller (Amersham, now GE Life Sciences).
10. Magnetic stirrer.
11. TBE buffer 10× (TBE 10× per liter:108 g Tris base, 55 g boric acid, 40 mL 0.5 M EDTA pH 8).
12. MDE acrylamide gel (Lonza, double strength stock solution) (*see* **Note 6**).
13. *N*,*N*,*N*′,*N*′,-tetramethylethylenediamine (TEMED).
14. 40 % ammonium persulfate (APS, wt/vol).
15. 100 mL beaker, vacuum flask, and magnetic flee.
16. Bind-silane solution [17 μL PlusOne Bind-Silane (Amersham, now GE life Sciences), 170 μL acetic acid 10 % (vol/vol), 10 mL ethanol].
17. PlusOne Repel-Silane (Amersham, now GE life Sciences).
18. Gel loading tips 0.2 mm (e.g., Starlab).

2.4 Preparing the ssDNA

1. Heating block for 1.5 mL microtubes (e.g., Eppendorf Thermomixer).
2. Denaturing dye solution (95 % formamide vol/vol, 10 mM NaOH, 0.002 % bromophenol blue and xylene cyanol wt/vol).
3. Ice bucket with crushed ice.

2.5 Band Visualization Type A

1. Trays made of stainless steel or plastic approx. 25×25 cm in pristine condition.
2. 10 % Acetic acid (vol/vol).
3. Deionized or distilled water.
4. Silver staining solution (per liter: 1 g silver nitrate, 1.5 mL 37 % formaldehyde).
5. Developing solution [per liter: 56.3 g sodium bicarbonate decahydrate, 2 mL 37 % formaldehyde, 1 mL sodium thiosulfate (0.2 % wt/vol)].

2.6 Band Visualization Type B

1. SybrGold (Invitrogen).
2. Visi-Blue transilluminator (UVP) or Dark Reader (Clare Chemical) or Storm scanner or Typhoon imager (both Amersham, now GE Life Sciences).

2.7 Extraction of ssDNA from Bands in SSCP Gels

1. Tray approx. 25×25 cm, deionized water.
2. Scalpel or razor blade, 1.5 mL micro reaction tubes.
3. DNA elution buffer [0.5 M ammonium acetate, 10 mM magnesium acetate, 1 mM EDTA pH 8, 0.1 % sodium dodecyl sulfate (SDS, wt/vol)].

4. Thermomixer (e.g. Eppendorf, for 1.5 mL tubes) or mixer for 1.5 mL tubes in 37 °C incubator.
5. Microtube centrifuge, refrigerated, 16,000 rcf minimum.
6. Ethanol (96 %).
7. Freezer compartment.
8. 10 mM Tris buffer (pH 8).

3 Methods

3.1 PCR and PCR Product Preparation

For SSCP, a 5′-end phosphorylated primer is needed. Several primers have been published for PCR-SSCP, most of which target the 16S rRNA gene [8, 10, 17]. This protocol refers to one of the so-called universal primer pairs, namely, COM1 and COM2-PH (the latter with a phosphate group at the 5′-end) covering the variable regions 4 and 5 of the 16S. PCR was performed with an initial denaturation at 94 °C for 3 min, followed by cycles of 1 min at 94 °C, 50 °C for 1 min, 72 °C for 1 min, and a final extension at 72 °C for 10 min. The number of cycles may vary according to primer and template choice. The COM primers have been found to be very efficient at an annealing temp of 50 °C and produce sufficient yields within 28–30 cycles. Amplification reaction components for 50 μL reaction volumes were as follows:

1. 1–1.5 U of hot start polymerase with supplied buffer.
2. 1.5–2.5 mM $MgCl_2$ (5 PRIME supplies buffer with 2.5 mM $MgCl_2$).
3. 0.5 μM Primers.
4. Deoxynucleotides, 200 μM each.
5. Optional: enhancer solution.
6. Approx. 1 ng template DNA or 0.1–1.0 μL of template DNA.

PCR products need to be further processed before they can be used for gel electrophoresis. The dsDNA PCR products are digested with lambda exonuclease (New England Biolabs, 5–10 U per 50 μL PCR) at 37 °C for 45 min in a total volume of 0.1 mL. In order to provide an optimal performance of the exonuclease, PCR products should be purified up front using a PCR purification kit. ssDNA products can be applied directly to the gel or can be purified again (PCR purification kit) to improve the fingerprint quality. The applications of PCR purification kits with a small elution volume of 10–12 μL are particularly recommended to maximize the amount of ssDNA that can be loaded onto the gel (*see* **Note 7**).

3.2 Gel Electrophoresis

For the SSCP gel electrophoresis, keeping a constant temperature is a key factor. Many acrylamide electrophoresis systems have only limited capability to maintain a constant temperature in the range of 0.1 °C. The Macrophor system is very efficient in controlling the gel temperature since gels are cast directly on a thermostatic plate connected to a cryostat/chiller.

A high voltage is necessary to migrate the ssDNA through the gel. Casting thin gels of approx. 0.4 mm is therefore recommended (*see* **Note 8**). The use of 0.6–0.65× MDE acrylamide gel (Lonza) has been found to deliver high-quality gel profiles. The following acrylamide gel concentrations apply for a 25 mL 20×20 cm gel, 0.4 mm thick:

1. 2.5 mL TBE stock solution (10×).
2. 7.8 mL MDE stock solution (2×).
3. 14.7 mL water.
4. 10 μL TEMED.
5. 25 μL 40 % APS.

Pouring the gel on a Macrophor system with 20 cm notched glass plates:

1. Clean glass and thermostatic plate with ethanol, and treat the thermostatic plate with repel-silane and the glass plate (optional) with binding silane (0.5 mL each) (*see* **Note 9**).
2. Position the thermostatic plate horizontally on the gel casting stand, and use levelling feet and a spirit leveller to ensure the system is level.
3. Fix 0.4 mm thick and 40 cm long spacers on both sides of the thermostatic plate with clamps and place the glass plate with the notches facing forward onto the lower part of the spacer (bind-silane solution side facing downwards).
4. To degas the gel, pour the gel matrix into a vacuum flask, add a magnetic stirrer, and apply a vacuum (water pump or membrane pump) for approx. 3 min to the solution so that a few air bubbles accumulate on the magnetic stirrer while stirring the matrix slowly on a magnetic platform.
5. Add APS and TEMED to the matrix solution and after a brief stir pour the matrix slowly onto the thermostatic plate and at the same time slowly slide the glass plate with the notches facing forward towards the upper location of the thermostatic plate.
6. Fix the glass plate to the thermostatic plate with clamps holding the spacers and insert the comb between the notched side of the glass plate and the thermostatic plate. Ensure that the gel casting system is level.
7. The acrylamide needs to polymerize for approx. 2 h at room temperature.

8. Fit the gel sandwiched between the glass plate and the thermostatic plate into the electrophoresis unit (glass plate facing the buffer chambers) and fill upper and lower buffer chamber with 1 L of 1× TBE buffer each.
9. Connect the thermostatic plate to the cryostat and set the temperature to 20 or 30 °C (*see* **Note 10**).

Preparing the ssDNA:

1. Mix 1 volume of purified ssDNA (5 μL or half of the PCR) with 1 volume (5 μL) of denaturing loading dye in a 1.5 mL reaction tube.
2. Heat the ssDNA sample to 95 °C to denature the single strands for 2 min in a heating block (*see* **Note 11**).
3. Cool the ssDNA on ice for 3 min before loading the gel.

Loading and running the gel:

1. Pull the comb and rinse the wells with a syringe and needle. For visualization purposes, load small volumes of denaturing dye into the wells and apply the voltage to the system for about 1 min, then rinse the wells again.
2. Load 10 μL of denatured sample into the well using a gel loading tip.
3. Apply 350–400 V and 8–10 mA for 16 h (6,000 V h).
4. After removing the thermostatic plate with the gel and glass plate attached, carefully lift the glass plate from the thermostatic plate without breaking the notches. The gel should stick to the glass plate. If you intend to use the gel for a subsequent DNA transfer, avoid using bind-silane solution on the glass plate (use a dedicated untreated glass plate for this type of experiment).

3.3 Band Visualization

1. Silver staining. For permanent visualization and cutting out of bands, silver staining of the gel has been proven to be most effective (Fig. 1a). Silver staining procedure:
   (a) Use two clean trays of stainless steel or plastic, and incubate the gel on the glass plate in 10 % acetic acid for 30 min.
   (b) Remove the acetic acid and wash the gel twice in deionized water for 5 min.
   (c) Remove the water and add silver staining solution and incubate in darkness for 30 min. The gel can be gently stirred but must not fall dry.
   (d) Remove the staining solution and rinse the gel briefly with deionized water (10 s).
   (e) Wash the tray with the gel in it in a small volume of cold (approx. 8 °C) developing solution for approx. 20 s.

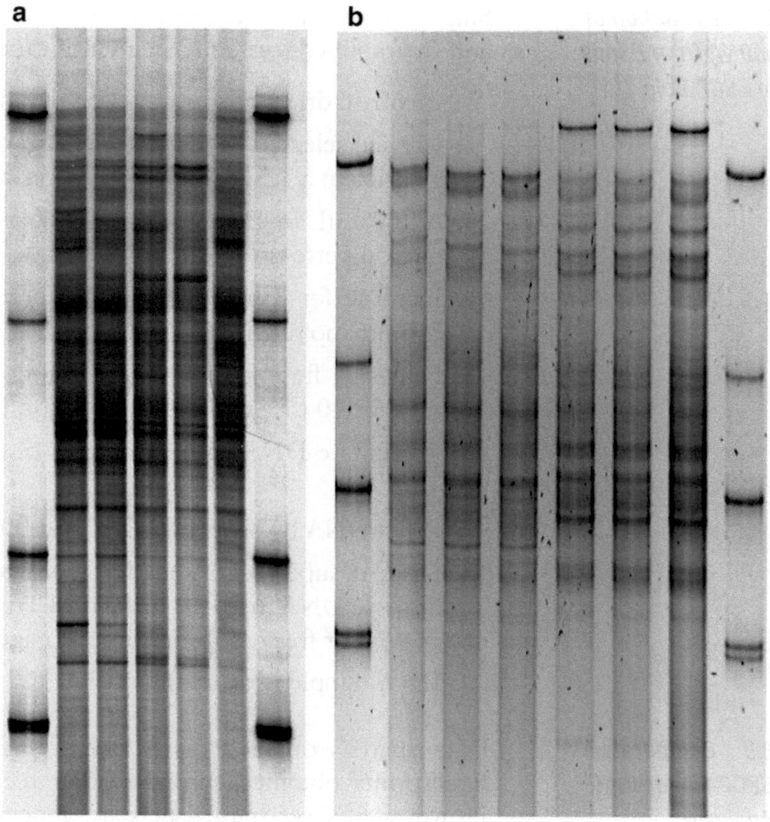

Fig. 1 SSCP fingerprints of various bacterial soil communities flanked by species standards (*Bacillus licheniformis*, *Rhizobium trifolii*, *Flavobacterium johnsoniae*, *Rhizobium radiobacter*) using a (**a**) Macrophor chamber (Amersham) and (**b**) Mutation Detection Chamber (CBS Scientific) in a walk-in cold room, 20 cm gel length at 20 °C

(f) Stain the gel in cold developing solution in the second tray until the bands are clearly visible.

(g) Stop the reaction by placing the gel into 10 % acetic solution for at least 10 min.

(h) Rinse the gel in deionized water and dry the gel.

2. SybrGold staining. Different stains can be employed to visualize ssDNA bands. A convenient form is to use SybrGold by overlaying the horizontal positioned gel for 30 min with 40 mL of 10,000× diluted SybrGold (in 1× TBE) in darkness. After submerging the gel in deionized water for 5 min, visualize the bands of the fingerprint (ssDNA) on a blue light or UV transilluminator (Fig. 1b) or a laser scanner/imager (Storm or Typhoon).

3.4 Extraction of ssDNA from Bands in SSCP Gels

Single bands can be cut out of dried and silver-stained SSCP gels and re-amplified for sequence identification (*see* **Note 12**).

1. Incubate dried gels in tray with deionized water for 5 min.
2. Cut out selected bands with a scalpel or razor blade and deposit gel slice in a 1.5 mL micro reaction tube.
3. Add 50 μL of DNA elution buffer and crush gel slice with a micropipette tip on the tube wall.
4. Incubate for 3 h at 37 °C in a Thermomixer with an orbit of 3–6mm, no rcf given.
5. Pellet gel fragments by centrifuging for 1 min 6,000 rcf and transfer 40 μL gel slice free solution to a new 1.5 mL tube.
6. Precipitate DNA with 2 volumes of ethanol (96 %) for 2 h at −20 °C.
7. Pellet DNA by centrifuging at 16,000 rcf for 15 min at 4–8 °C.
8. Discard supernatant, air-dry DNA pellet for 5–10 min, and dissolve DNA in 10 μL of 10 mM Tris buffer (pH 8), and store at 8 °C or freeze for longtime storage.
9. DNA sample is ready for PCR.

3.5 Comparative SSCP Fingerprint Analysis

Data analysis of SSCP gels is very similar to image analysis of fingerprints obtained through other techniques such as DGGE. However, if silver-stained gels are generated, then a scanner with an integrated top light is highly recommended to obtain digitalized fingerprints. We have used GelCompar (Applied Maths) and Phoretix (Nonlinear Dynamics) software to analyze SSCP gels, but other software packages may also be suitable. These software packages were used to normalize the fingerprints and to carry out the cluster analysis embedded in the software, e.g., Unweighted Pair Group Method with Arithmetic Mean (UPGMA). Alternatively, normalized fingerprints can be exported to carry out Principal Component Analysis (PCA) or Correspondence Analysis (CA) with software packages such as Canoco (Microcomputer Power).

4 Notes

1. Although a variety of polymerases can be employed to obtain a PCR product for SSCP analysis, the use of hot start polymerases such as HotMaster (5 PRIME), Platinum (Invitrogen), or Robust HotStart (Kapa Biosystems) Taq improved the yield and reduced the amount of unspecific products.
2. Self-made PCR water: Filter ultrapure (18 MΩ) water through a 0.2 μm membrane filter into sterile 1.5 mL safe lock microtubes and autoclave tubes locked, filled with 1 mL water each.
3. Universal primers are published for SSCP applications targeting the variable regions V2–3, V4–5, and V6–8 of the 16S [13, 17].

In addition, Peters et al. [9] demonstrated the diversity analysis of fungi by SSCP. Bacterial groups were also targeted more specifically employing a nested PCR approach [10].

4. DNA obtained from DNA extraction kits such as FastDNA spin kit for soil (MP Biomedicals) or Ultra Clean Soil DNA kit (MO BIO Laboratories) can be used directly or five times diluted as template DNA for the PCR.

5. Several other electrophoresis systems can be used for running SSCP gels. (1) the Protean XL from Biorad (comes with a cooling core that can be connected to a chiller); (2) the Scie-Plas 20×20 cm protein chamber with an integrated temperature exchange chamber (now replaced by the TV400 series) has been used successfully by connecting the latter to a chiller and inserting magnetic stirrers; (3) the TGGE maxi system (Biometra) has also been used successfully to run SSCP gels. The system allows keeping a constant temperature. Unfortunately, horizontal gels have to be run in this chamber which can cause problems when running thin gels for 16 h. (4) The DCode system (Biorad) can also be used with an optional cooling coil, but the standard coil size is too small to obtain sufficient constant temperatures; (5) in some cases where systems can only heat but not cool, the placing of the electrophoresis unit in a 4 °C walk-in room can be a solution, and this has been done successfully with a Mutation Detection Chamber from CBS Scientific (Fig. 1b).

6. Standard acrylamide is sufficient to cast an SSCP gel, but the MDE gel solution has been used most successfully to generate high-quality fingerprints and is therefore recommended.

7. The lambda exonuclease digest can be carried out with raw PCR since the HotMaster buffer (5 PRIME) shows similar features to the lambda exonuclease buffer. However, optimal results will be obtained by purifying the PCR product up front and using the buffer supplied with the exonuclease. The digested product can be applied to the gel directly but then the buffer solution in the digest will create smiling effects in the gel. For optimal presentation, use a PCR purification kit with small elution volume such as the Apache from Invitek, the PCR purification kit from Fermentas, or the Qiagen PCR MinElute. By eluting into 10–12 μL, one can apply 50 % of a PCR in a single well.

8. Running SSCP gels that are 1 mm thick can cause problems since the voltage will be significantly lower at 10 mA and new settings have to be tried. With the TGGE maxi, it is impossible to reach the 400 V, and the electrophoresis has to be carried out at a lower voltage setting or the strength of TBE has to be reduced. Many small standard power packs only allow a voltage of 200 or 300 V. Users have to source power packs that can run at a higher voltage.

9. Using bind-silane solution and repel-silane was found to be very useful when using the Macrophor system to run SSCP gels. Bind-silane solution is beneficial when SSCP gels are stained with silver nitrate as many wash steps are included in the staining procedure. However, bind-silane solution has to be avoided if the ssDNA in the gel needs to be transferred to a membrane. The use of GelBond films with the hydrophobic side facing the gel side is recommended although gels can also be transferred from the untreated glass plate to a sheet of blotting paper by rubbing the blotting paper over the gel and then peeling it off carefully. For the application of electroblotting and Southern blot hybridization, please refer to Schmalenberger and Tebbe [14].
10. SSCP running conditions have been optimized to 1× TBE buffer at 20 °C but can also be used to run gels at 10–30 °C. We have used different temperatures for different primers to obtain sharp bands in the fingerprints, and users who try new primers are advised to optimize for the best temperature. Liu and Sommer [18] patented the SSCP5 method to detect mutations using multiple temperature settings. However, if an apparatus is used that cannot cool actively, it is advised to run the gels at the highest possible temperature without losing a ssDNA conformation effect.
11. Different additives have been tried for the loading/denaturing buffer and the SSCP gel but the gel recipes above showed the best results. As the lambda exonuclease digests dsDNA to single-stranded DNA, SSCP gels can be loaded without denaturation, but in practice the denaturation step improved the quality of the fingerprint.
12. Retrieving ssDNA from SSCP bands was most successful by using silver-stained gels. DNA in the gels was found to be stable for several years. It is also possible to retrieve ssDNA from SybrGold-stained gels, but the exposure to UV light is usually disintegrating the DNA fragments quickly, and we recommend the use of a blue light transilluminator. However, faint bands can be easily overlooked by this method due to low-signal intensity.

Acknowledgements

This work was supported by the British Natural Environment Research Council (NERC).

References

1. Liu WT et al (1997) Characterization of microbial diversity by determining terminal restriction fragment length polymorphisms of genes encoding 16S rRNA. Appl Environ Microbiol 63:4516–4522
2. Muyzer G, de Waal EC, Uitterlinden AG (1993) Profiling of complex microbial populations by denaturing gradient gel electrophoresis analysis of polymerase chain reaction-amplified genes coding for 16S rRNA. Appl Environ Microbiol 59:695–700
3. Muyzer G, Smalla K (1998) Application of denaturing gradient gel electrophoresis (DGGE) and temperature gradient gel electrophoresis (TGGE) in microbial ecology. Antonie Leeuwenhoek 73:127–141
4. Orita M et al (1989) Detection of polymorphisms of human DNA by gel-electrophoresis as single-strand conformation polymorphisms. Proc Natl Acad Sci U S A 86:2766–2770
5. Schwieger F, Tebbe CC (1998) A new approach to utilize PCR-single-strand-conformation polymorphism for 16S rRNA gene-based microbial community analysis. Appl Environ Microbiol 64:4870–4876
6. Hayashi K (1992) PCR-SSCP - a method for detection of mutations. Genet Anal Biomol Eng 9:73–79
7. Lee DH, Zo YG, Kim SJ (1996) Nonradioactive method to study genetic profiles of natural bacterial communities by PCR-single-strand-conformation polymorphism. Appl Environ Microbiol 62:3112–3120
8. Tebbe CC et al (2001) Single-strand conformation polymorphism (SSCP) for microbial community analysis. In: Rochelle PA (ed) Environmental molecular microbiology: protocols and applications. Horizon Scientific Press, Wymondham, pp 161–175
9. Peters S et al (2000) Succession of microbial communities during hot composting as detected by PCR-single-strand-conformation polymorphism-based genetic profiles of small-subunit rRNA genes. Appl Environ Microbiol 66:930–936
10. Dohrmann AB, Tebbe CC (2005) Effect of elevated tropospheric ozone on the structure of bacterial communities inhabiting the rhizosphere of herbaceous plants native to Germany. Appl Environ Microbiol 71:7750–7758
11. Atha DH et al (1998) Detection of p53 point mutations by single strand conformation polymorphism: analysis by capillary electrophoresis. Electrophoresis 19:172–179
12. King S, McCord BR, Riefler RG (2005) Capillary electrophoresis single-strand conformation polymorphism analysis for monitoring soil bacteria. J Microbiol Methods 60: 83–92
13. Schmalenberger A, Tebbe CC (2002) Bacterial community composition in the rhizosphere of a transgenic, herbicide-resistant maize (*Zea mays*) and comparison to its non-transgenic cultivar Bosphore. FEMS Microbiol Ecol 40:29–37
14. Schmalenberger A, Tebbe CC (2003) Bacterial diversity in maize rhizospheres: conclusions on the use of genetic profiles based on PCR-amplified partial small subunit rRNA genes in ecological studies. Mol Ecol 12:251–261
15. Fracchia L et al (2006) Bacterial diversity in a finished compost and vermicompost – differences revealed by cultivation-independent analyses of PCR-amplified 16S rRNA genes. Appl Microbiol Biotechnol 71:942–952
16. Schmalenberger A et al (2008) Two-dimensional single strand conformation polymorphism (SSCP) of 16S rRNA gene fragments reveals highly dissimilar bacterial communities in an acidic fen. Eur J Soil Biol 44:495–500
17. Schmalenberger A, Schwieger F, Tebbe CC (2001) Effect of primers hybridizing to Different evolutionarily conserved regions of the small-subunit rRNA gene in PCR-based microbial community analyses and genetic profiling. Appl Environ Microbiol 67:3557–3563
18. Liu Q, Sommer SS (2001) Multi-conditional SSCP (SSCP5): a rapid method for mutation scanning with virtually 100% sensitivity. Patent US, City of Hope, USA

Chapter 7

Human Fecal Source Identification with Real-Time Quantitative PCR

Orin C. Shanks, Lindsay Peed, Mano Sivaganesan, Richard A. Haugland, and Eunice C. Chern

Abstract

Waterborne diseases represent a significant public health risk worldwide and can originate from contact with water contaminated with human fecal material. We describe a real-time quantitative PCR (qPCR) method that targets a genetic marker of the human-associated *Bacteroides dorei* for identification of human fecal pollution in ambient water samples. The following protocol includes water sample collection, filtration, DNA isolation with a sample processing control, qPCR amplification with an internal amplification control, and quality control data analysis.

Key words Real-time quantitative PCR (qPCR), Fecal source identification, Human fecal pollution, Absolute quantification

1 Introduction

Recreational waters are one of the world's most valued resources. Fecal waste from human sources and other animal sources can contaminate surface waters and pose a serious threat not only to the environment, but also to human health and coastal water community economies. Traditional methods to identify fecal pollution in impaired waters rely on the cultivation of fecal indicator bacteria such as *Escherichia coli* and enterococci [1]. Elevated levels of these microorganisms alert local water quality managers to the presence of fecal pollution. However, these methods offer no information as to the originating sources of fecal pollution, which is a critical piece of information needed to design effective mitigation strategies and monitor their effectiveness.

The application of the real-time quantitative polymerase chain reaction (qPCR) technology to the development of human-associated fecal source identification methods is revolutionizing the way water quality managers characterize pollution and make decisions.

Ian T. Paulsen and Andrew J. Holmes (eds.), *Environmental Microbiology: Methods and Protocols*, Methods in Molecular Biology, vol. 1096, DOI 10.1007/978-1-62703-712-9_7, qPCR allows for the continuous monitoring of a particular genetic marker as amplification occurs. Estimation of the concentration of the genetic marker is based on the theoretical premise that there is a log-linear relationship between the starting amount of DNA target in a reaction and the fractional thermal cycle where qPCR product accumulation is first significantly detectable [2]. qPCR methods designed to estimate human-associated fecal source identification genetic marker concentrations are gaining widespread attention [3–17].

We describe a qPCR method that targets a *Bacteroides dorei* human-associated genetic marker for the identification of human fecal pollution in ambient water samples [3, 17, 18]. The following protocol includes water sample collection, filtration, DNA isolation with a SPC (SPC), qPCR amplification with an internal amplification control (IAC), and quality control data analysis.

2 Materials

2.1 Water Sample Collection and Filtration

1. 500 mL sterile sample collection bottles.
2. Disposable filtration unit.
3. Membrane filters: sterile, polycarbonate, 0.4 μm pore size, 47 mm in diameter (*see* **Note 1**).

2.2 DNA Isolation from Filter

1. Preloaded bead extraction tubes: 2 mL O-ring screw cap semi-conical microcentrifuge tubes containing 0.3 g of acid-washed glass beads.
2. Salmon testis DNA: 10 mg/mL.
3. AE buffer: 10 mM Tris-Cl, 0.5 mM EDTA, pH 9.0.
4. DNA-EZ silica column kit (GeneRite Inc, North Brunswick, NJ) (*see* **Note 2**).

2.3 Real-Time Quantitative PCR

1. PCR grade/UltraPure water.
2. AE buffer: 10 mM Tris-Cl, 0.5 mM EDTA, pH 9.0.
3. Taqman® Fast Universal Master Mix (Applied Biosystem, Foster City, CA).
4. Bovine serum albumin (BSA): 2 mg/mL, fraction V powder.
5. Plasmid DNA standards in the following concentrations: 10^1–10^6 copies 2 μL^{-1}.
6. 96-well FAST qPCR reaction plate (Applied Biosystem, Foster City, CA).
7. Aluminum adhesive qPCR plate covers.
8. Primers and hydrolysis probes (*see* Table 1) (*see* **Note 3**).

Table 1
Primer and hydrolysis probe sequences for HF183 and Sketa22 qPCR assays

	Sequence (5′–3′)	Reference
HF183 qPCR assay		
HF183	ATCATGAGTTCACATGTCCG	[3]
BFDRev	CGTAGGAGTTTGGACCGTGT	[18]
BFDFAM	(FAM) CTGAGAGGAAGGTCCCCCACATTGGA (TAMRA)	
UC1P1	(VIC) CCTGCCGTCTCGTGCTCCTCA(TAMRA)	[17, 20]
Sketa22 qPCR assay		
SketaF2	GGTTTCCGCAGCTGGG	[17]
SketaR2	CCGAGCCGTCCTGGTC	
SketaP	(FAM) AGTCGCAGGCGGCCACCGT (TAMRA)	

The BFDFAM and SketaP hydrolysis probes are 5′ labeled with FAM reporter dye and 3′ labeled with a TAMRA quencher. The HF183 qPCR assay IAC hydrolysis probe is 5′ labeled with VIC reporter dye and 3′ labeled with a TAMRA quencher. Rehydrate primers and hydrolysis probes with AE buffer to make a stock solution of 500 μM of each primer and 100 μM of the hydrolysis probe. Make a primers and hydrolysis probe mixed working solution by adding 10 μL of each primer and 4 μL of hydrolysis probe to 576 μL of UltraPure water for a total volume of 600 μL. Store stock and working solutions at −20 °C

2.4 Laboratory Organization and Dedicated Equipment

Contamination from extraneous sources potentially introduced throughout a qPCR method can be problematic. DNA from equipment, other samples, and previously synthesized amplicons can contaminate qPCR amplifications leading to false positives and misinterpretation of results. Extraneous DNA from these sources can be limited through the use of physical barriers and dedicated equipment. It is recommended that sample filtering, DNA isolation, qPCR reagent assembly, and qPCR amplifications occur in four separate laboratories with dedicated equipment. In addition to physical barriers and dedicated equipment, qPCR analysis should progress in a single direction (Fig. 1). Unidirectional progression prevents backtracking of purified DNA from environmental and reference samples, as well as qPCR amplicons generated from DNA amplification.

2.4.1 Sample Filtration

1. Vacuum filtration manifold.
2. Forceps: sterilize between use by dipping into 95 % ethanol and flaming tips of forceps.
3. Disposable gloves.

2.4.2 DNA Isolation/Template Addition

1. Disposable gloves.
2. Pipettors (p200 and p1000) with barrier tips for DNA isolation.
3. Bead beater.
4. Microcentrifuge.
5. 1.7 mL low-retention microtubes.
6. Dedicated p20 pipette with barrier tips for template addition.

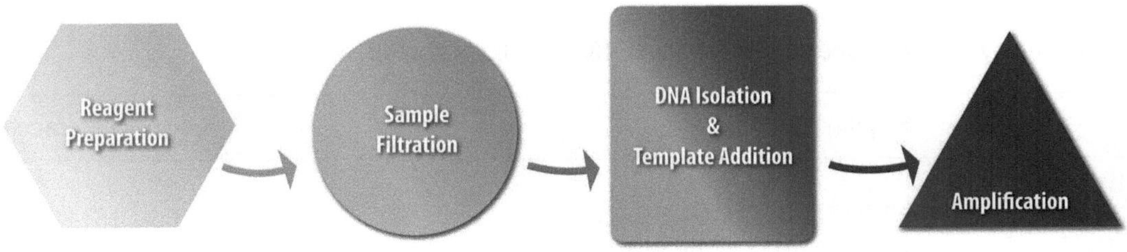

Fig. 1 Recommended physical separation and unidirectional progression of analysis for HF183 qPCR method

2.4.3 qPCR Reagent Mixing

1. Disposable gloves.
2. DNA-free refrigerator and freezer for reagents, primers, and hydrolysis probes.
3. Laminar flow hood or PCR work station.
4. Pipettors (p10, p200, and p1000).
5. Vortex mixer.
6. Optical FAST 96-well PCR reaction tray (Applied Biosystem, Foster City, CA).
7. Optical adhesive PCR reaction tray tape (Applied Biosystem, Foster City, CA).

2.4.4 Genetic Marker Amplification

1. Model 7900 HT sequence detector (Applied Biosystem, Foster City, CA).
2. 96-well centrifuge.

3 Methods

3.1 Preparation of Plasmid-Derived DNA Standards

1. Plasmid-derived DNA standards contain a sequence corresponding to the HF183 primers and the BFDFAM hydrolysis probe (Table 1).
2. Plasmid includes binding sites for forward and reverse assay primers, as well as the BFDFAM hybridization sequence (Fig. 2) and can be constructed in-house or ordered from companies specializing in custom gene synthesis (*see* **Note 4**).
3. Digest plasmid with appropriate restriction endonuclease. Ideal restriction site should result in a single cut to linearize plasmid and be situated at least 100 base pairs up- or downstream of the standard DNA insert.
4. Clean digested product using a DNA clean-up kit such as QiaQuick PCR purification kit (Qiagen).
5. Measure spectrophotometric absorbance of cleaned product at 260 nm (A_{260}) in triplicate and average the readings.

Fig. 2 Plasmid-derived calibration curve standard construct design. *Arrows* indicate forward and reverse primer locations. *Line* with *closed circles* represents location of BFDFAM hydrolysis probe

6. Use plasmid size to determine plasmid copies per gram as follows (where *X* indicates the total number of base pairs in the plasmid including the standard DNA construct insert):

$$\frac{6.023\times10^{23}\,molecules\,/\,g}{(X\,bp)(615\,Da\,/\,bp)} = molecules\,/\,g \qquad (1)$$

7. Use absorbance reading to calculate the number of the plasmid copies in 2 μL. Use this value to make the following plasmid dilutions: 10^6, 10^5, 10^4, 10^3, 10^2, and 10^1 copies/μL.
8. Prepare aliquots of each dilution and store in low-retention plastic microcentrifuge tubes at −20 °C. Aliquots should be discarded after three freeze/thaw cycles to minimize the effect of template degradation.

3.2 Preparation of Sample Processing Control (SPC)

1. Dilute salmon DNA to 1 mg/mL by adding 0.5 mL of the 10 mg/mL stock to 5 mL AE buffer.
2. Make a 10 μg/mL salmon DNA solution by adding 80 μL of the 1 mg/mL salmon DNA solution to 8 mL of AE buffer.
3. Measure spectrophotometric absorbance of the 10 μg/mL salmon DNA solution at 260 nm (A_{260}).
4. Make 1 mL aliquots of the 10 μg/mL solution and store in low-retention plastic microcentrifuge tubes at 4 °C (*see* **Note 5**).
5. Use absorbance reading (1 OD = 50 μg/mL) to calculate volume needed to make 0.2 μg/mL salmon DNA working stock.

$$\frac{(A_{260}value)\times(50\mu g\,/\,ml)}{0.2\mu g\,/\,ml} = required\ volume \qquad (2)$$

6. Dilute 10 μg/mL solution to 0.2 μg/mL according to calculation.
7. Store salmon DNA working stock solution (0.2 μg/mL) in low-retention plastic microcentrifuge tube at 4 °C (*see* **Note 6**).

3.3 Preparation of Internal Amplification Control (IAC)

1. Internal amplification control plasmids contain a sequence corresponding to the HF183 primers and a universal probe sequence (Table 1; Fig. 3).

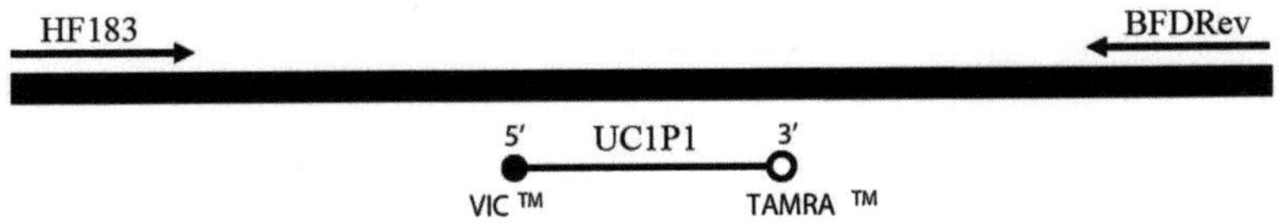

Fig. 3 IAC standard construct design. *Arrows* indicate forward and reverse primer locations. *Line* with *closed circles* represents location of UC1P1 hydrolysis probe

2. Plasmids containing the sequence being measured can be constructed in-house or ordered from vendors specializing in custom gene synthesis (*see* **Note 4**).
3. Digest plasmid with appropriate restriction endonuclease. Ideal restriction site should result in a single cut to linearize plasmid and be situated at least 100 base pairs up- or downstream of IAC insert.
4. Clean digested product using a purification kit such as QiaQuick PCR purification kit (Qiagen).
5. Measure spectrophotometric absorbance of cleaned product at 260 nm (A_{260}) in triplicate and average the readings.
6. Use plasmid size to determine plasmid copies per gram as follows (where X indicates the total number of base pairs in the plasmid including the standard DNA construct insert):

$$\frac{6.023\times10^{23}\,molecules\,/\,\mathrm{g}}{(X\,bp)(615\,Da\,/\,bp)}=molecules\,/\,\mathrm{g} \qquad (3)$$

7. Use absorbance reading to calculate the plasmid concentration. Use this value to make the following plasmid dilution: 25 copies/μL.
8. Prepare aliquots of this dilution and store in low-retention plastic microtube at −20 °C. Aliquots should be discarded after three freeze/thaw cycles to minimize the effect of template degradation on results.

3.4 Water Sample Collection and Filtration

1. Collect 500 mL of water in sterilized sample collection bottle. Transport back to laboratory on ice for immediate filtration.
2. Place disposable filter unit on vacuum filtration manifold (*see* **Note 7**).
3. Prepare a filter blank by measuring 100 mL of PCR grade water and transferring into the filter funnel (*see* **Note 8**).
4. Vacuum filter the water through the polycarbonate membrane filter.
5. Place filter into a 2 mL preloaded bead tube by folding the filter on the filter base into a cylinder with sample side facing

inward using sterile forceps. After folding the membrane, dip forceps in alcohol and flame to sterilize.

6. Place a new disposable filter unit onto the vacuum manifold.
7. Gently shake environmental water sample to uniformly suspend bacteria.
8. Measure and vacuum filter 100 mL of environmental water sample through the polycarbonate membrane filter.
9. Rinse the sides of the filter funnel with approximately 20 mL of PCR grade water.
10. Place filter into a 2 mL preloaded bead tube as described in **step 5**.
11. Repeat **steps 6–10** to obtain triplicate filters for each environmental sample.
12. Once all environmental water samples are filtered in triplicate, prepare an additional filter blank (PCR grade water only). *See* **steps 2–5**.
13. Store bead tubes containing folded filters at −80 °C until time of DNA isolation (*see* **Note 9**).

3.5 DNA Isolation from Filter

1. Add 600 μL of AE buffer containing 0.2 μg/mL salmon DNA working stock to each bead tube (contains folded filter and silica beads). In addition, prepare three extraction blanks (AE buffer containing salmon DNA only) and at minimum of one blank without salmon DNA (omit salmon DNA in AE buffer) per batch of samples extracted.
2. Bead beat extraction tubes at 6 m s^{-1} for 60 s.
3. Centrifuge bead tubes in microcentrifuge at 14,000 × *g* for 3 min.
4. Carefully transfer supernatant to a sterile 1.7 mL low-retention plastic microcentrifuge tube without disturbing the pellet.
5. Centrifuge microcentrifuge tubes containing supernatant at 14,000 × *g* for 1 min to remove any remaining debris.
6. Transfer 400 μL supernatant to a second sterile 1.7 mL low-retention microcentrifuge tube. It is important to pipette exact amount to yield reliable SPC data.
7. Purify DNA extracts by adding 1 mL (or twice the volume of supernatant) of binding buffer to the clarified supernatant and mix gently by pipetting.
8. Place columns from DNA-EZ kit into elution tubes.
9. Transfer approximately 750 μL of the mixture to the column and centrifuge at 12,000 × *g* for 1 min.
10. Discard flow through and repeat **step 9** using the remaining mixture and the same column.

Table 2
Preparation of primer and hydrolysis probe mix for HF183 multiplex qPCR assay

Reagent	Stock solution (μM)	Volume in primer/ probe mix (μL)
HF183	500	10
BFDRev	500	10
FAM probe (BFDFAM)	100	4
VIC probe (UC1P1)	100	4
UltraPure H_2O		572

11. Add 500 μL of wash buffer into the column.
12. Centrifuge at 12,000×*g* for 1 min. Discard flow through and repeat **step 11**.
13. Place the column into a sterile 1.7 mL low-retention plastic microcentrifuge tube.
14. Add 75 μL of warm (60 °C) elution buffer into the column and centrifuge tube at 12,000×*g* for 1 min.
15. Repeat **step 14** for a total elution volume of 150 μL.
16. Transfer to a 200 μL low-retention plastic microtube.
17. Measure spectrophotometric absorbance of purified DNA at 260 nm (A_{260}) in triplicate and average the readings.
18. Store purified DNA at 4 °C. The qPCR analysis should be performed within 48 h of extraction.

3.6 HF183 Multiplex Real-Time Quantitative PCR

1. Prepare primers and hydrolysis probe mix as follows (*see* Table 2).
2. Calculate volume of reagents needed for qPCR analysis (Table 3). When analyzing multiple samples, make a master mix of reagents and aliquot appropriate volumes of the master mix into individual wells. For example, if there are 96 samples to be analyzed, include extra reaction volume into calculation and prepare reagents as follows in Table 3.
3. Aliquot 23 μL of mix into each well of a 96-well FAST qPCR reaction plate.
4. Lightly cover plate with aluminum adhesive qPCR tape (do not seal tape onto plate).
5. Label and store on ice in the dark for transport to dedicated laminar flow hood in DNA isolation laboratory for the addition of DNA template.
6. Add 2 μL of DNA template into appropriate wells using a dedicated pipette. Include a minimum of three no-template controls

Table 3
Preparation of master mix for HF183 multiplex qPCR assay

Reagent	Final concentration	Volume for one 25 μL reaction (μL)	Example: 106 reactions (μL)
Taqman fast master mix	1×	12.5	1,325
BSA	0.2 mg/mL	2.5	265
Primer/probe mix	1 μM/80 nM	3.5	371
UltraPure H_2O	–	3.5[a]	477
IAC plasmid	–	1	106
DNA template		2[a]	

[a] *See* **Note 10**

(2 μL of PCR grade water substituted for DNA template), two extraction blanks, and two positive controls (prepared from known human fecal source material) on each plate. Analyze environmental water samples in duplicate (*see* **Note 11**).

7. In addition, add 2 μL of each plasmid DNA standard in triplicate for each concentration ranging from 10 to 1×10^6 copies on each plate (*see* **Note 12**).
8. Seal plate with optical adhesive PCR tape.
9. Place plate into real-time qPCR instrument according to manufacturer's instructions. Set amplification conditions for HF183 qPCR assay at 50 °C for 2 min followed by 95 °C for 10 min and then forty cycles of 95 °C for 15 s and 60 °C for 1 min. Analyze data with the cycle quantification (C_q) set at 0.03 for Applied Biosystem instruments.

3.7 Sketa22 Real-Time Quantitative PCR

The protocol for the Sketa22 qPCR assay follows the HF183 qPCR assay protocol (*see* Subheading 3.6) with a few exceptions: use of the Sketa22 primers and hydrolysis probe in place of the IAC hydrolysis probe and plasmid and HF183 primers and hydrolysis probe. Prepare the primer and hydrolysis probe mix and master mix as described in Tables 4 and 5.

3.8 Data Quality Control Parameters

Quality control parameters are necessary to generate reliable estimates of genetic marker concentration in unknown samples. Errors can arise from numerous sources in the qPCR method protocol ranging from improper sample handling, degradation of DNA standards, and laboratory technician difficulties to interferences originating from the unknown sample itself. As a result, a series of recommended control protocols are used to help ensure the generation of high-quality qPCR data.

Table 4
Preparation of primer and hydrolysis probe mix for Sketa22 qPCR assay

Reagent	Stock solution (μM)	Volume in primer/probe mix (μL)
SketaF2	500	10
SketaR2	500	10
FAM probe (SketaP)	100	4
UltraPure H_2O		576

Table 5
Preparation of master mix for Sketa22 qPCR assay

Reagent	Final concentration	Volume for one 25 μL reaction (μL)	Example: 106 reactions (μL)
Taqman fast master mix	1×	12.5	1,325
BSA	0.2 mg/mL	2.5	265
Primer/probe mix	1 μM/80 nM	3.5	371
UltraPure H_2O	–	4.5[a]	477
DNA template		2[a]	

[a] *See* **Note 10**

3.8.1 Evaluation of Calibration Curve

Two criteria used to determine the suitability of the calibration curve for transforming raw quantification cycle (C_q) data into sample concentration estimates include amplification efficiency (E) and the coefficient of determination (R^2). E is derived from the slope parameter in the fitted curve and is defined as follows:

$$E = \left(10^{-1/slope}\right) - 1 \tag{4}$$

It is recommended that an E value should be within the range of 0.90–1.10. R^2 is the proportion of variability in the DNA standard C_q measurements that is accounted for by the regression model and it can range from 0 to 1. A $R^2 \geq 0.90$ is recommended for applying a calibration curve for estimating unknown sample concentrations.

3.8.2 Evaluation of Sample Processing Efficiency

Loss of DNA, interference from the environmental matrix, and laboratory technician error can all impact the efficiency of DNA recovery during extraction from an environmental sample filter. To monitor for variability in sample processing efficiency, each

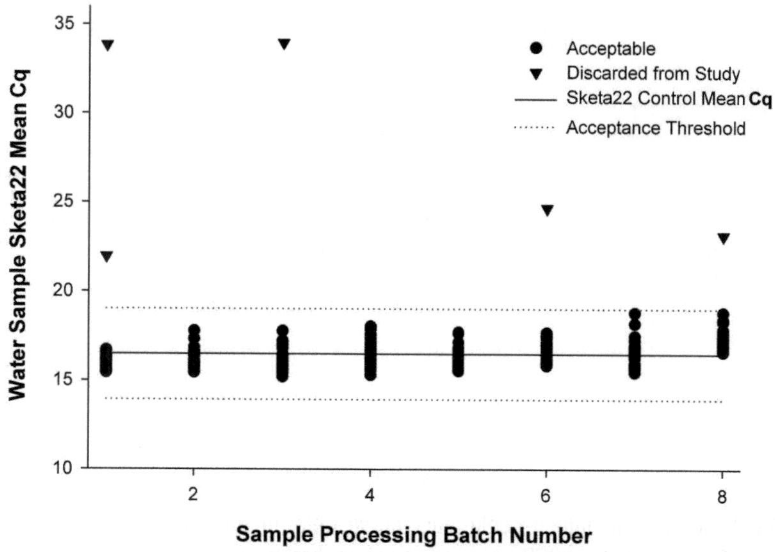

Fig. 4 Interpretation of SPC data. The *solid line* denotes the Sketa22 control mean C_q. The *dotted line* indicates the acceptance threshold range based on the Sketa22 control mean ± 3 C_q. *Circles* represent environmental samples with acceptable sample processing efficiency. *Triangles* represent samples that fail SPC metric and should be discarded from future data interpretation

environmental sample filter is spiked with a fixed concentration of salmon sperm DNA, and the resulting DNA elute is tested with the Sketa22 qPCR assay. The demonstration of consistent recovery efficiency from one sample DNA extract to the next is achieved by establishing an acceptance threshold based on repeated control experiments. The sample processing efficiency threshold should be based on a minimum of three extraction blanks containing salmon DNA spike (triplicate C_q measurements for each extraction blank DNA extract) per extraction batch. A mean is then calculated from resulting extraction blank Sketa22 qPCR C_q data. A commonly used acceptance threshold range is $\pm 3C_q$ of the established extraction blank control mean C_q (Fig. 4). For DNA extracts that elicit Sketa22 qPCR C_q values outside the acceptance threshold, associated HF183 qPCR C_q should be discarded from the data set.

3.8.3 Screening Unknown Samples for Amplification Interference

Substances inhibitory to qPCR amplification can persist after DNA purification. Therefore, an IAC designed to evaluate the suitability of isolated DNA for qPCR-based amplification should be performed with each environmental sample DNA extract. The criterion for concluding no significant qPCR amplification interference can be established as a mean $C_q \pm 2$ standard deviations based on repeated experiments across several instrument runs (minimum of 30 replicates) where 25 copies of the HF183 IAC is mixed with buffer only and amplified. Mean (μ) and standard deviation (σ) are

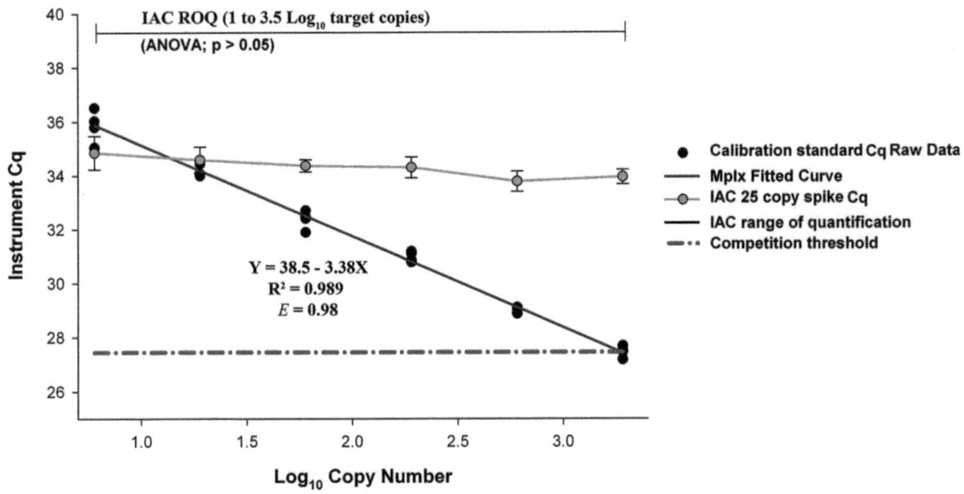

Fig. 5 Plotting the multiplex calibration curve, determination of IAC range of quantification, and establishing the HF183/IAC multiplex qPCR assay competition threshold

then estimated from resulting C_q data. Average of all the C_q values gives an estimate for μ. To account for between batch variability and within batch variability, a one-way analysis of variance (ANOVA) is used to estimate the σ, and the estimate is given by

$$\hat{\sigma} = \sqrt{\hat{\sigma}_b^2 + \hat{\sigma}_e^2} \quad (5)$$

where $\hat{\sigma}_b^2$ and $\hat{\sigma}_e^2$ are the respective estimates for between and within batch variances [19].

Environmental sample IAC C_q data can then be reviewed to determine if respective IAC Cq values are within the acceptable range of variability. IAC C_q values greater than the upper bound acceptance threshold can result from either amplification inhibition or competition between the IAC and native HF183 DNA targets in the multiplex HF183/IAC qPCR reaction.

In order to establish a target competition threshold, users must generate a multiplex calibration curve, determine the IAC range of quantification (ROQ), and establish a threshold for HF183/IAC competition (Fig. 5). The IAC ROQ is determined by performing an analysis of variance (ANOVA) test comparing raw C_q measurements from the plasmid DNA standards used to generate the multiplex calibration curve. The range of plasmid DNA standard concentrations where there is no significant difference in C_q measurements starting from lowest DNA standard concentration to the highest represents the IAC ROQ. The competition threshold is the C_q value where the upper bound of the IAC ROQ intersects the multiplex HF183 calibration curve. For example, Fig. 5 depicts an IAC ROQ range from $\log_{10}$ 0.5–3.5 plasmid DNA standard concentrations. The competition threshold is then determined to be where the upper bound of the IAC ROQ ($\log_{10}$ 3.5 copy number) intersects the multiplex calibration curve (27.5 C_q).

The combination of establishing an interference threshold based on buffer-only IAC spike C_q measurements and a competition threshold derived from the IAC ROQ and HF183 multiplex calibration curve provides the tools to differentiate between IAC inhibition (IAC C_q > upper bound interference threshold and > competition threshold) and competition (IAC C_q > upper bound interference threshold, but < competition threshold). It is recommended that data associated with a particular DNA extract where IAC C_q values are observed outside the accepted range are either discarded from the study or the DNA extract is diluted and retested.

3.8.4 Detection of Extraneous DNA

To monitor for potential false positives, a series of controls (filter blanks, extraction blanks, and no-template amplification controls) are included throughout the qPCR method. These controls consist of laboratory grade water instead of environmental water samples and should be included with each sample filtration event, DNA isolation step, and thermal cycle instrument run as indicated above. All false positives observed over the course of study should be reported.

3.8.5 Evaluation of Positive Controls

Positive controls of a known concentration for the HF183 and Sketa22 qPCR assays should be included with each thermal cycle instrument run (*see* **Note 13**). Controls should generate C_q values within two standard deviations of the mean generated from a series of repeated control experiments.

3.9 Estimation of Genetic Marker Concentration in Environmental Sample

The protocol described above utilizes an absolute standard curve approach. A calibration (fitted) curve can be generated using a variety of different mathematical approaches ranging from simple linear regression to Bayesian statistics [20]. For most studies, it is recommended that data for a calibration curve are generated for each thermal cycle instrument run. However, some study designs can allow for the use of other strategies (*see* for review 21). Regardless of the approach used (*see* **Note 14**), calibration curves should include five or six known DNA concentrations with triplicate C_q measurements at each standard concentration.

4 Notes

1. We recommend using either the 0.4 μm pore size (47 mm in diameter) polycarbonate membrane filter (GE Osmonics, Minnetonka, MA) or the Super-200 polyethersulfone 0.2 μm pore size (47 mm in diameter) membrane filter (PALL, Ann Arbor, MI).
2. We suggest the use of this kit for extraction and purification of DNA as we found it provides optimal yields with little inhibition.

However, other kits may be used if they also provide desirable results.

3. Since qPCR requires a much higher concentration of primers than traditional PCR, we suggest purchasing primers in high concentration.
4. We suggest sequencing the insert to confirm that the primer and hydrolysis probe sequences are correct.
5. We recommend preparing fresh salmon DNA solution at least every 6 months.
6. Prepare the working stock salmon DNA fresh daily when needed.
7. Depending on the disposable filtration unit used, the membrane filter included may need to be removed and replaced with either a 0.2 or 0.4 μm pore size (47 mm in diameter) polycarbonate or polyethersulfone membrane filter.
8. We recommend filtering at least three filter blanks for each sample filtering event.
9. Filters can be stored at −80 °C for a maximum of 12 months before DNA isolation.
10. Volumes may change depending on specifics of individual experiment.
11. Although duplicate analysis is described here, we recommend increasing the number of replicates analyzed to provide better precision of results.
12. Current protocol recommends generating a standard curve with each instrument run. For studies requiring more than six instrument runs, it may be beneficial to use a master standard curve approach produced by a minimum of six standard curves in separate instrument runs.
13. The range of positive control concentrations should encompass the concentration expected in unknown samples.
14. We recommend that the calibration curve strategy selected account for variability when possible.

Acknowledgement

The US Environmental Protection Agency, through its Office of Research and Development, funded and managed the research described herein. It has been subjected to the Agency's peer and administrative review and has been approved for external publication. Any opinions expressed in this paper are those of the author(s) and do not necessarily reflect the official positions and policies of the US EPA. Any mention of trade names or commercial products does not constitute endorsement or recommendation for use.

References

1. USEPA (1986) Ambient water quality criteria; availability. Fed Regist 51:8012–8016
2. Sambrook J, Russell DW (2001) Molecular cloning: a laboratory manual, vol 1–3, 4th edn. Cold Spring Harbor Laboratory Press, New York, NY
3. Bernhard AE, Field KG (2000) A PCR assay to discriminate human and ruminant feces on the basis of host differences in *Bacteroides-Prevotella* genes encoding for 16S rRNA. Appl Environ Microbiol 66:4571–4574
4. Caldwell JM, Raley ME, Levine JF (2007) Mitochondrial multiplex real-time PCR as a source tracking method in fecal-contaminated effluents. Environ Sci Technol 41:3277–3283
5. Carson CE et al (2005) Specificity of *Bacteroides thetaiotaomicron* marker for human feces. Appl Environ Microbiol 71:4945–4949
6. Dombek PE et al (2000) Use of repetitive DNA sequences and PCR to differentiate *Escherichia coli* isolates from human and animal sources. Appl Environ Microbiol 66:2572–2577
7. Kildare BJ et al (2007) 16S rRNA-based assays for quantitative detection of universal, human-, cow-, and dog-specific fecal *Bacteroidales*: a Bayesian approach. Water Res 41:3701–3715
8. Layton A et al (2006) Development of *Bacteroides* 16S rRNA gene TaqMan-based real-time PCR assays for estimation of total, human, and bovine fecal Pollution in water. Appl Environ Microbiol 72:4214–4224
9. McQuaig SM et al (2006) Detection of human-derived fecal pollution in environmental waters by use of a PCR-based human polyomavirus assay. Appl Environ Microbiol 72: 7567–7574
10. Okabe S et al (2007) Quantification of host-specific *Bacteroides-Prevotella* 16S rRNA genetic markers for assessment of fecal pollution in freshwater. Appl Microbiol Biotechnol 74:890–901
11. Pina S et al (1998) Viral pollution in the environment and in shellfish: human adenovirus detection by PCR as an index of human viruses. Appl Environ Microbiol 64:3376–3382
12. Scott TM et al (2005) Potential use of a host associated molecular marker in *Enterococcus faecium* as an index of human fecal pollution. Environ Sci Technol 39:283–287
13. Seurinck S et al (2005) Detection and quantification of human-specific HF183 *Bacteroides* 16S rRNA genetic marker with real-time PCR for assessment of human faecal pollution in freshwater. Environ Microbiol 7:249–259
14. Shanks OC et al (2009) Quantitative PCR for genetic markers of human fecal pollution. Appl Environ Microbiol 75:5507–5513
15. Ufnar JA et al (2006) Detection of the *nifH* gene of *Methanobrevibacter smithii*: a potential tool to identify sewage pollution in recreational waters. J Appl Bacteriol 101:44–52
16. Lee CS, Lee J (2010) Evaluation of new *gyrB-based* real-time PCR system for the detection of *B. fragilis* as an indicator of human-specific fecal contamination. J Microbiol Meth 82: 311–318
17. Haugland RA et al (2010) Evaluation of genetic markers from the 16S rRNA gene V2 region for use in quantitative detection of selected *Bacteroidales* species and human fecal waste by real-time PCR. Syst Appl Microbiol 33:348–357
18. Converse RR et al (2009) Rapid QPCR-based assay for fecal *Bacteroides* spp. as a tool for assessing fecal contamination in recreational waters. Water Res 43:4828–4837
19. Montgomery DC (1991) Design and analysis of experiments, 3rd edn. Wiley, New York, NY
20. Sivaganesan M et al (2008) A Bayesian method for calculating real-time quantitative PCR calibration curves using absolute plasmid DNA standards. BMC Bioinformatics 9:120
21. Sivaganesan M et al (2010) Improved strategies and optimization of calibration models for real-time PCR absolute quantification. Water Res 44:4726–4735

Chapter 8

Next Generation Barcode Tagged Sequencing for Monitoring Microbial Community Dynamics

Katy Breakwell, Sasha G. Tetu, and Liam D.H. Elbourne

Abstract

Microbial identification using 16S rDNA variable regions has become increasingly popular over the past decade. The application of next-generation amplicon sequencing to these regions allows microbial communities to be sequenced in far greater depth than previous techniques, as well as allowing for the identification of unculturable or rare organisms within a sample. Multiplexing can be used to sequence multiple samples in tandem through the use of sample-specific identification sequences which are attached to each amplicon, making this a cost-effective method for large-scale microbial identification experiments.

Key words Microbial identification, 16S rDNA, 454, Amplicon sequencing, Multiplex sequencing

1 Introduction

Early microbial identification required the isolation of an organism in pure culture, followed by a combination of morphological observation, physiological and biochemical tests. This greatly limited the study of microbial diversity as it is estimated that over 99 % of all microorganisms cannot be cultivated using traditional culture-based techniques [1]. The process of microbial identification has developed rapidly over the past several decades, moving from phenotypic observation to DNA sequencing-based methods. The most popular locus currently targeted is the 16S ribosomal RNA gene. This gene, also known as the small subunit (SSU) ribosomal gene, is a phylogenetic marker widely used for microbial identification and phylogenetic classification. It is present within all known prokaryotic organisms and consists of a homologous backbone interspersed with nine regions of hypervariability (Fig. 1). These regions of hypervariability have been used for the identification of organisms to the species or strain level [2–4].

The advent of next generation sequencing technologies has further advanced the field of microbial identification, overcoming limitations experienced with both traditional techniques and

Ian T. Paulsen and Andrew J. Holmes (eds.), *Environmental Microbiology: Methods and Protocols*, Methods in Molecular Biology, vol. 1096, DOI 10.1007/978-1-62703-712-9_8, © Springer Science+Business Media, LLC 2014

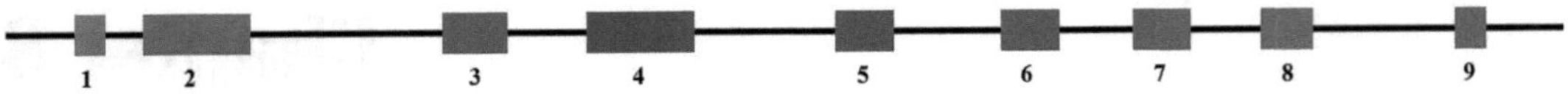

Fig. 1 16S rRNA gene. A schematic of the 16S rRNA gene depicting highly conserved regions (*black*) and hypervariable regions (1–9)

Sanger sequencing, and is capable of surveying samples in far greater depth than ever before [5]. It allows for the identification of rare and unculturable organisms, as well as the sequencing of almost entire microbial ecosystems [3, 6, 7]. It can be further applied to microbial identification studies through large-scale sequencing of 16S rDNA.

Amplicon sequencing involves the large-scale sequencing of a particular region of DNA (e.g., 16S rDNA). The target DNA is amplified using PCR, and the products are processed and sequenced using next generation sequencing. Amplicon sequencing is used in preference to metagenomic sequencing for microbial identification, as only DNA of interest is sequenced. This vastly increases the amount of relevant data returned compared with metagenomic sequencing, resulting in a far greater depth of information. In addition, multiple samples can be sequenced in tandem using multiplex sequencing. In multiplex sequencing, each sample is given a unique identification tag, generally attached to the 5′ end of the DNA. The DNA from each sample is pooled and sequenced, resulting in a large amount of sequence data from multiple samples. Sequences can then be assigned to the original samples according to the identification tags.

The following is a representative protocol for multiplex amplicon sequencing of 16S rDNA on the 454 sequencing platform. Multiplex sequencing is also available on the Solexa/Illumina sequencing platform.

2 Materials

2.1 454 Multiplex Sequencing

1. FastStart high fidelity PCR system (Roche).
2. QIAquick PCR purification kit (QIAGEN).
3. NanoDrop 2000 (Thermo Scientific).

2.2 Bioinformatic Processing and Analysis

The following list contains some frequently used software for microbial community analysis. This is not an exhaustive list as software is frequently being advanced and developed (*see* **Note 1**). All of the software used is open source and either freely available for download (as source, executables, or both for a wide variety of operating systems) or accessible via a web interface.

1. The Ribosomal Database Project (RDP) is a web-based software suite that can currently be used for microbial identification,

sequence processing, and some single sample analysis [8]. It is available at http://rdp.cme.msu.edu/.

2. Mothur is a comprehensive and highly flexible software suite that can currently be used for microbial identification, sequence processing, and single sample analysis and comparative sample analysis [9]. It is available at http://www.mothur.org/wiki/Download_mothur and comprehensive instructions are available from http://www.mothur.org/wiki/Main_Page.
3. Neighbor is a program written by Joseph Felsenstein and provided by the University of Washington. It is used to infer neighbor-joining phylogenetic trees [10]. It is available for use as part of the PHYLIP suite available for download at http://evolution.genetics.washington.edu/phylip.html.

3 Methods

3.1 454 Multiplex Sequencing

3.1.1 Fusion Primer Design

Fusion primers should be designed to include a Lib-L primer A sequence (Roche), a sample identification sequence and the target primer sequence within the sequencing primer, and a Lib-L primer B sequence (Roche) followed by the target primer sequence within the reverse primer (Fig. 2).

Lib-L Primer A and B

The fusion primer set must be designed to include Lib-L primer A and B (Roche). The Lib-L primer A sequence must be located on the sequencing primer (*see* **Note 2**).

Sample Identification Sequence

The sample identification sequence is a molecular barcode unique to each sample included in a multiplexed sequencing run. Sample identification sequences must be included in the sequencing primer. 454 Life Sciences provides premade multiplex identifiers (MIDs), as well as MID sequences, that can be utilized during library preparation or fusion primer design. Alternatively, sample identification sequences can be designed and included in the fusion primer sequence (*see* **Note 3**).

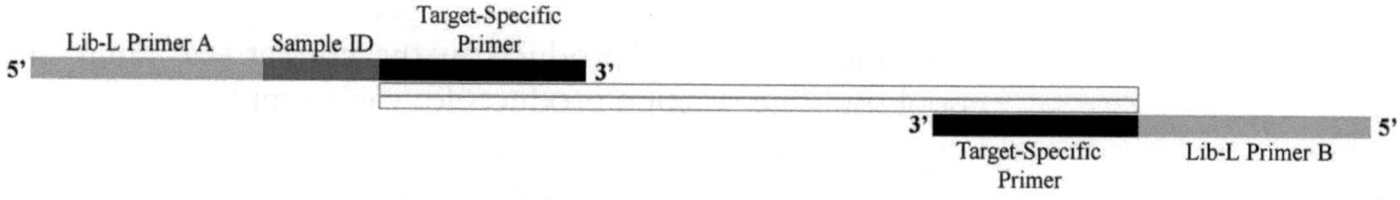

Fig. 2 Fusion primer set. A schematic of a fusion primer set bound to target DNA. Each fusion primer consists of a target-specific sequence which includes the Lib-L primer A or B primer sequence at the 5′ end of the target-specific sequence. A sample-specific ID should be included between the Lib-L primer A sequence and the target-specific sequence for multiplex sequencing

Target Primer Sequence

The target primer sequences bind directly to your sample DNA during amplification (*see* **Note 4**).

3.1.2 Amplicon Production

1. Amplify DNA using the fusion primers in a gradient thermocycler with the annealing temperature spanning 10 °C at approximately 0.8 °C intervals, centered on the primer set melting temperature.
2. Visualize the PCR products using gel electrophoresis. The lowest annealing temperature to produce a band of the appropriate size is considered the optimal annealing temperature (*see* **Note 5**). If the optimal annealing temperature is not generated using a 10 °C temperature range, the amplification can be repeated using an expanded range of up to 20 °C. Optimization should be carried out for all fusion primer sets on the appropriate samples.
3. For each sample, perform three identical amplification reactions using the following conditions (*see* **Note 6**):

FastStart high fidelity reaction buffer	5 μL
Template	2 μL
Primer	1 μL of each primer (50 μM)
PCR grade nucleotide mix	1 μL (200 μM of each dNTP)
FastStart high fidelity DNA polymerase	0.5 μL
Total volume	50 μL

Initial denaturation	95 °C for 2 min
Denaturation	95 °C for 30 s
Annealing	Optimal temperature as determined above for 30 s
Extension	72 °C for 60 s
PCR cycles	35
Final extension	72 °C for 5 min

4. Visualize the PCR products using gel electrophoresis to confirm the presence of products in the correct size range and pool triplicate reaction products for each sample.
5. Use the QIAquick PCR purification kit, or other DNA purification method, to purify the PCR products.
6. Calibrate the NanoDrop 2000 using PCR grade water and blank using the elution liquid from **step 5**. For each sample, load 1–2 μL of purified PCR product onto the NanoDrop

2000 stage and measure using the default nucleotide settings. Samples are generally required to contain at least 500 ng of DNA for library preparation and sequencing. If the total amount of DNA in a sample is less than 500 ng, perform one or more additional PCRs, clean up, and pool all reactions for that sample. The purity of a sample is measured by comparing the 260 nm/280 nm absorbance ratio. Samples should have a 260 nm/280 nm ratio of greater than 1.8. If the 260 nm/280 nm ratio is below 1.8, repeat **steps 5** and **6**.

7. Send prepared products to a 454 sequencing center for analysis. Sequencing centers may require samples to be quantified and pooled prior to library preparation. In this case, it is advised to use an assay, such as the PicoGreen® dsDNA quantitation assay (Invitrogen), for quantification. When pooling samples after quantification, an equal amount of DNA from each sample should be used.

3.2 Bioinformatic Analysis

3.2.1 Deconvolution

Individual reads can be deconvoluted using the sample identification sequence. Depending on the program used to deconvolute reads, each sequence will be assigned to a group based on its sample identification sequence, with an additional group containing sequences that lack a sample identification sequence. Sequences that lack a sample identification sequence must be removed from the analysis, as they cannot be reliably assigned to any sample.

3.2.2 Quality Processing

Sequences are processed to remove primer and sample identification sequences as they do not reflect the gene sequence of the organism and often confound further analysis. Sequences can also be processed to remove low-quality sequences prior to further analysis. Some factors which may be used to identify low-quality sequences include:

1. Primer presence: The presence of both the forward and reverse primers within a sequence indicates that the full target sequence was obtained. Sequences that lack a forward or reverse primer indicate that sequencing was terminated prior to the complete sequencing of the amplicon. These terminated sequences can be considered to be low quality. This is only applicable if the length of the target gene region does not exceed the maximum length of sequence reads.
2. QUAL file: Sequences which have an average quality score of below 20 have an average base call accuracy of below 99 % and can therefore be considered low quality.
3. Length: Sequences that are much smaller or much larger than the target gene region can be artifacts produced during amplification.

4. Ambiguous nucleotides: Ambiguous nucleotides, e.g., N, indicate low quality as the information received during sequencing was insufficient to make a specific base call.
5. Homopolymer regions: Sequences containing long homopolymer regions are a known artifact associated with 454 sequencing. They can be a result of one strong base signal read as multiple base signals, resulting in an incorrect string of identical bases in the sequence in place of a single base.
6. Chimeric sequences: Chimeric sequences contain sequence from two or more template sequences. These should be removed as they provide incorrect sequence information.

3.2.3 Community Analysis

There are multiple methods that can be used to analyze microbial community data. The first steps involved in analysis sort sequences into groups based on genetic distance, known as operational taxonomic units (OTUs; *see* **Note 7**). To convert sequence data to OTU data, sequences must be aligned, a pairwise distance matrix calculated, and that information used to cluster sequences into OTUs. Software to perform these steps is available from both the RDP website and the Mothur program.

1. Alignment: An alignment is created by arranging the position of bases in each sample sequence to reflect its position in the gene. The output is a single alignment file with sequences aligned to a reference sequence (*see* **Note 8**).
2. Pairwise Distance Matrix: A pairwise distance matrix is created by calculating the genetic distance between every sequence within the alignment file.
3. Cluster: A cluster file is created by using the genetic distance information from the pairwise distance matrix to sort the sequences into groups. The data is generally sorted into groups at a range of distances, from unique to identical.

Once data have been clustered, there are a wide range of options for community analysis. The analyses performed will depend on the research question that is being asked. Below is a brief outline of some of the options available:

1. Classification: The identity of organisms within a sample can be obtained by comparing the 16S rDNA sequences from that sample with 16S rDNA sequences of known organisms. Software for this analysis is available from the RDP website or using the Mothur software suite.
2. Rank abundance curves: Rank abundance curves are a method of visually displaying the diversity of a sample. The curves are created by arranging OTUs from most abundant to least abundant along the *x*-axis and plotting the abundance, usually in numbers of sequences, on the *y*-axis. The number of OTUs in

a sample will give an indication of the diversity, while the abundance of each OTU will show the community composition. Software for this analysis is available from the RDP website or using the Mothur software suite.

3. Rarefaction curves: Rarefaction curves give a species richness estimate that can be used to compare multiple samples. They are produced by taking random subsets of a sample and plotting the cumulative number of novel species or OTUs obtained against the cumulative total number of sequences. A steep curve indicates that there is unsampled diversity in the population, as each new random subset contains a large number of novel species or OTUs. A curve that becomes parallel to the *x*-axis indicates that most of the population diversity was sampled, as there are few novel species or OTUs introduced with each random subset. Software for this analysis is available from the RDP website or using the Mothur software suite.
4. Venn diagram: Venn diagrams are a method of visually displaying the number of species or OTUs unique to each sample and shared between each sample. It allows for a comparison of community composition between multiple samples. Software for this analysis is available using the Mothur software suite.
5. UniFrac: UniFrac is an algorithm that uses a combined neighbor-joining tree to carry out pairwise comparisons of community structure between all samples. It measures the total fraction of unique branch length contributed by each sample to the tree to calculate a UniFrac score. This will determine whether any sample has a significantly different structure to other samples in the analysis. A neighbor-joining tree can be created using Neighbor program, while software for UniFrac analysis is available using the Mothur software suite.

4 Notes

1. The software available for microbial community analysis is constantly changing as current programs become outdated and new programs and methods are developed. Examining additional software options prior to analysis is advised. Some additional software that is currently available includes:

   Greengenes [11], an online database and set of tools for analysis of 16S rDNA sequences (http://greengenes.lbl.gov/cgi-bin/nph-index.cgi)

   The Bioinformatics Toolkit, downloadable software developed by K. E. Ashelford and provided by the School of Biosciences, Cardiff University (http://www.bioinformatics-toolkit.org/index.html)

MG-RAST [12], an online tool for metagenomic analysis with some functions that can be utilized for 16S rDNA analysis (http://metagenomics.anl.gov/).

2. The Lib-L primer A sequence binds the amplicon to the sequencing beads during library preparation. The amplicon will then be sequenced from the end that contains the Lib-L primer A sequence. If both the forward and reverse primers contain Lib-L primer A, then both ends of the amplicon will bind to the bead, preventing clonal PCR and sequencing of that amplicon. If both the forward and reverse primers contain Lib-L primer B, neither end of the amplicon will bind to the sequencing bead, and the amplicon will be washed away prior to clonal PCR and sequencing. Therefore, it is important that the Lib-L primer A sequence be present within the sequencing fusion primer, while the Lib-L primer B sequence be present within the reverse fusion primer.
3. It is recommended that sample IDs are between 6 and 20 base pairs in length. Sample IDs less than 6 base pairs in length risk the identification sequence being removed during sequencing due to low-quality base call. Sample IDs greater than 20 base pairs in size will increase the size of the fusion primer, increasing the likelihood of nonspecific sequences being generated during amplicon production. If you design your own sample identification tags, it is best to avoid including homopolymer regions (e.g., AAA or TTT) as these can be problematic during 454 sequencing. It is also recommended that each sample ID be sufficiently different from all others that a single base change will not prevent identification.
4. The region amplified will depend on several factors, including the specificity of information required, the target organism(s), and the variable region chosen for amplification. Selecting the most useful region of the 16S rDNA gene is perhaps the most important step in microbial identification experiments, as there is no single hypervariable region that can be used for identification of all eubacteria [13]. In addition, there are few primers available that reliably amplify both eubacteria and archaea. The depth of identification information should also be considered, as more specific information may require amplification of a longer region of the 16S rDNA gene. Another consideration is the sequencing platform that will be used. Ideally, the amplicon length should be close to the maximum reliable sequence length generated by the sequencing platform to be used.
5. 16S rDNA PCR will sometimes generate several closely spaced bands due to differences in hypervariable region size between bacteria. Therefore, when optimizing the PCR using fusion primers, the optimal annealing temperature may produce several closely spaced bands around the expected product size.

6. Where possible, it is good practice to obtain and analyze biological replicates for all samples. Inclusion of biological replicates will give an approximation of the biological variance of the sample, and therefore more robust conclusions can be drawn. It is also good practice to include technical replicates where possible. If it is not possible to treat each technical replicate as a separate sample, several technical replicates should be produced and pooled before sequencing to reduce variation due to sample preparation [14].
7. Operational taxonomic units (OTUs) are often used in large-scale microbial studies because of the high proportion of novel sequences encountered, as the vast majority of microorganisms are not systematically classified. An OTU is a group of sequences that are within a specific genetic distance of each other and can be used to describe groups at any taxonomic rank, from species to domain. A distance of 0.03 is commonly used to group eubacterial species equivalents, while a distance of 0.05 is commonly used to form eubacterial genus equivalent groups.
8. It is advantageous to use alignment software that takes in to account the secondary structure of 16S rDNA. This is because each nucleotide in the structure will have a different selection pressure depending on its location within the gene. Hypervariable regions experience less selective pressure, as these regions are linking the structural regions.

References

1. Amann R, Ludwig W, Schleifer KH (1995) Phylogenetic identification and in situ detection of individual microbial cells without cultivation. Microbiol Rev 59:143–169
2. Tringe SG, Hugenholtz P (2008) A renaissance for the pioneering 16S rRNA gene. Curr Opin Microbiol 11:442–446
3. Sogin ML, Morrison HG, Huber JA, Mark Welch D, Huse SM, Neal PR, Arrieta JM, Herndl GJ (2006) Microbial diversity in the deep sea and the underexplored "rare biosphere". Proc Natl Acad Sci U S A 103: 12115–12120
4. Huber JA, Mark Welch DB, Morrison HG, Huse SM, Neal PR, Butterfield DA, Sogin ML (2007) Microbial population structures in the deep marine biosphere. Science 318:97–100
5. Schloss PD, Handelsman J (2005) Introducing DOTUR, a computer program for defining operational taxonomic units and estimating species richness. Appl Environ Microbiol 71:1501–1506
6. Edwards RA, Rodriguez-Brito B, Wegley L, Haynes M, Breitbart M, Peterson DM, Saar MO, Alexander S, Alexander EC Jr, Rohwer F (2006) Using pyrosequencing to shed light on deep mine microbial ecology. BMC Genomics 7:57
7. Tringe SG, von Mering C, Kobayashi A, Salamov AA, Chen K, Chang HW, Podar M, Short JM, Mathur EJ, Detter JC, Bork P, Hugenholtz P, Rubin EM (2005) Comparative metagenomics of microbial communities. Science 308:554–557
8. Cole JR, Wang Q, Cardenas E, Fish J, Chai B, Farris RJ, Kulam-Syed-Mohideen AS, McGarrell DM, Marsh T, Garrity GM, Tiedje JM (2009) The ribosomal database project: improved alignments and new tools for rRNA analysis. Nucleic Acids Res 37:D141–D145
9. Schloss PD, Westcott SL, Ryabin T, Hall JR, Hartmann M, Hollister EB, Lesniewski RA, Oakley BB, Parks DH, Robinson CJ, Sahl JW, Stres B, Thallinger GG, Van Horn DJ, Weber CF (2009) Introducing mothur: open-source, platform-independent, community-supported software for describing and comparing microbial communities. Appl Environ Microbiol 75:7537–7541
10. Felsenstein J (1989) PHYLIP - phylogeny inference package (Version 3.2). Cladistics 5: 164–166

11. DeSantis TZ, Hugenholtz P, Larsen N, Rojas M, Brodie EL, Keller K, Huber T, Dalevi D, Hu P, Andersen GL (2006) Greengenes, a chimera-checked 16S rRNA gene database and workbench compatible with ARB. Appl Environ Microbiol 72:5069–5072
12. Meyer F, Paarmann D, D'Souza M, Olson R, Glass EM, Kubal M, Paczian T, Rodriguez A, Stevens R, Wilke A, Wilkening J, Edwards RA (2008) The metagenomics RAST server—a public resource for the automatic phylogenetic and functional analysis of metagenomes. BMC Bioinformatics 9:386
13. Chakravorty S, Helb D, Burday M, Connell N, Alland D (2007) A detailed analysis of 16S ribosomal RNA gene segments for the diagnosis of pathogenic bacteria. J Microbiol Methods 69:330–339
14. Prosser JI (2010) Replicate or lie: the need for replication. Environ Microbiol 12:1806–1810

Chapter 9

Analysis of Methanotroph Community Structure Using a *pmoA*-Based Microarray

Guy C.J. Abell, Nancy Stralis-Pavese, Yao Pan, and Levente Bodrossy

Abstract

The analysis of methanotroph community composition is relevant to studies of methane oxidation in a number of environments where methane is a significant carbon source. The development and application of a microarray targeting the particulate methane monooxygenase gene (*pmoA*) have allowed a high-throughput, semiquantitative analysis of the major methanotroph groups in a number of different environments.

Here we describe the use of a *pmoA*-based short oligo array for the analysis of methanotroph populations in sediment samples. The method is suitable for analysis of any type of environmental sample from which DNA can be extracted.

Key words Microarray, Methanotroph, *pmoA*, Hybridization

1 Introduction

The analysis of functional diversity in the environment is an important step towards understanding microbial ecology. The application of microarray techniques to microbial ecology has improved our ability to describe the diversity of microorganisms in the environment. Microarrays are well suited to microbial ecology, providing high sample throughput and highly parallel detection of complex microbial communities in a wide range of samples. Microarrays allow semiquantitative characterization of target genes by means of specific hybridization of labelled target gene sequences, amplified from the environment, to the corresponding oligonucleotide probes on a small solid surface (glass slide). The functional genes targeted in this way can be involved in any number of ecological and environmental processes of interest, such as nitrogen or methane metabolism.

A comprehensive microarray comprising short oligonucleotide probes (17–28 nt long) complementary to the particulate methane monooxygenase (*pmoA*) and closely related ammonia

Ian T. Paulsen and Andrew J. Holmes (eds.), *Environmental Microbiology: Methods and Protocols*, Methods in Molecular Biology, vol. 1096, DOI 10.1007/978-1-62703-712-9_9, © Springer Science+Business Media, LLC 2014

monooxygenase (*amoA*) genes, encoding for one of the subunits of the particulate methane monooxygenase (pMMO) and the ammonia monooxygenase (AMO), respectively, has been designed [1, 2]. The pMMO catalyzes methane oxidation in methanotrophs, whereas the AMO catalyzes the oxidation of ammonia in nitrifiers. Both methane and ammonia oxidations are important environmental processes that are crucial to biogeochemical transformations of nitrogen and methane, as such they are linked to environmentally important processes such as productivity, global warming, soil fertility, and eutrophication.

The first step in the development of such a microarray is the design of oligonucleotide probes targeting the different groups of organisms represented by the *pmoA* gene. In silico designed probes are then validated using a comprehensive set of pure cultures and environmental clones covering almost the entire known diversity of methanotrophs and bacteria carrying *pmoA*.

pmoA/*amoA* genes are then PCR amplified from environmental samples or reference strains/clones (for validation) and used as template for a subsequent in vitro transcription reaction (IVT) to generate labelled single-stranded RNA transcripts (*see* also **Note 1**). Labelled targets are then fragmented and hybridized to complementary oligonucleotide probes on the array. Validation and data analysis have been previously described in detail by Bodrossy et al. and Stralis-Pavese et al. [1, 2], and an updated probe set is described in [3]. The method has been used for the study of a number of environments including landfill cover soil [1], alpine meadow soil [4], coal mine soil [5], estuarine sediment [6], and peat [7] and peat moss [8].

2 Materials

All solutions are prepared using milli-Q water and the pH adjusted prior to autoclaving. Solutions that are not sterilized by autoclaving are noted. All solutions used for working with RNA are RNAse-treated by incubation overnight at 37 °C with 1/1,000 DEPC prior to autoclaving.

2.1 Microarray Components

1. VSS-25 silyated (aldehyde) slides (TeleChem International, Inc., Sunnyvale, CA, USA).
2. Oligonucleotides, including a 5′ NH_2 group followed by a C_6 spacer and five thymidine residues preceding the probe sequence.
3. DMSO.
4. 0.2 % SDS (do not autoclave).
5. Sodium borohydrate solution: Dissolve 1.5 g $NaBH_4$ in 450 mL phosphate buffered saline (PBS) and then add 133 mL 100 % ethanol. This needs to be prepared in a fume hood, just prior to use.

2.2 DNA Extraction Components

1. FastDNA Spin kit for soil (MP Biomedicals, Solon, OH, USA), includes FastPrep tubes, MT Buffer, binding matrix, wash buffer, and spin filters.
2. Lysis Buffer: combine 39 mL of 200 mM NaH_2PO_4 and 61 mL of 200 mM Na_2HPO_4, add 17.54 g NaCl, 2 g CTAB, and 4 g PVP K30, and adjust to pH 7.0 and make up to 200 mL, autoclave. Just prior to use, add lysozyme to 5 mg/mL.
3. Phenol:chloroform:isoamyl alcohol (24:24:1) pH 7.8.
4. Chloroform:isoamyl alcohol (24:1).
5. Phase Lock gel, heavy 2 mL tubes (5Prime Inc., Gaithersburg, MD, USA).
6. Proteinase K: dissolve 10 mg of proteinase K in 1 mL of sterile dH_2O, can be stored frozen for up to 1 month.
7. 1.7 mL microcentrifuge tubes (sterile).

2.3 PCR and IVT Components

1. Invitrogen Taq polymerase kit (Invitrogen, USA) contains 10× PCR Buffer, 5 U/μL Taq polymerase, and 50 mM $MgCl_2$.
2. 10 mM DNTP mixture.
3. 200 μL PCR strip tubes with caps.
4. PCR primers:

   pmoA189F (GGNGACTGGGACTTCTGG), pmoA682R (GAASGCNGAGAAGAASGC), and mb661R-T7 (TAATACGACTCACTATAGCCGGMGCAACGTCYTTACC).
5. Agarose gel (1.5 % w/v) dissolved in TAE Buffer.
6. High Pure PCR Purification Kit (Roche, USA).
7. RNA polymerase kit (Invitrogen, USA) contains 5× RNA polymerase buffer, 100 mM DTT, and 40 U/μL RNA polymerase.
8. Solutions (10 mM) of each of ATP, CTP, GTP, and UTP.
9. 5 mM CY3-UTP (GE, USA).
10. RNAse-free water.
11. RNasin, 40 U/μL (Promega, USA).
12. RNeasy RNA extraction kit (QIAGEN, USA) contains RLT Buffer, RNeasy mini column, 2 mL collection tube, and RPE buffer.
13. 99 % Ethanol.
14. 1 M Tris–HCl pH 7.4.
15. 100 mM $ZnSO_4$.
16. 500 mM EDTA, pH 7.5.

2.4 Hybridization Components

1. RNAse-free water.
2. 10 % SDS (do not autoclave).
3. 50× Denhardt's reagent (Sigma, USA).

4. 20× SSC.
5. HybriWell triple chambers (Grace Bio-Labs, USA), include sealing spots.
6. 2× SSC, 0.1 % SDS.
7. 0.2× SSC.
8. 0.1× SSC.

3 Methods

All procedures are carried out at room temperature (25 °C) unless otherwise stated. Points where the procedure may be paused by storing reactions at −20 °C are noted (*pause point*). A detailed description of the methodology for database construction, probe design, array printing, and validation is given in reference [3]. An overview of the experimental procedure is given in Fig. 1. The general procedure is given below.

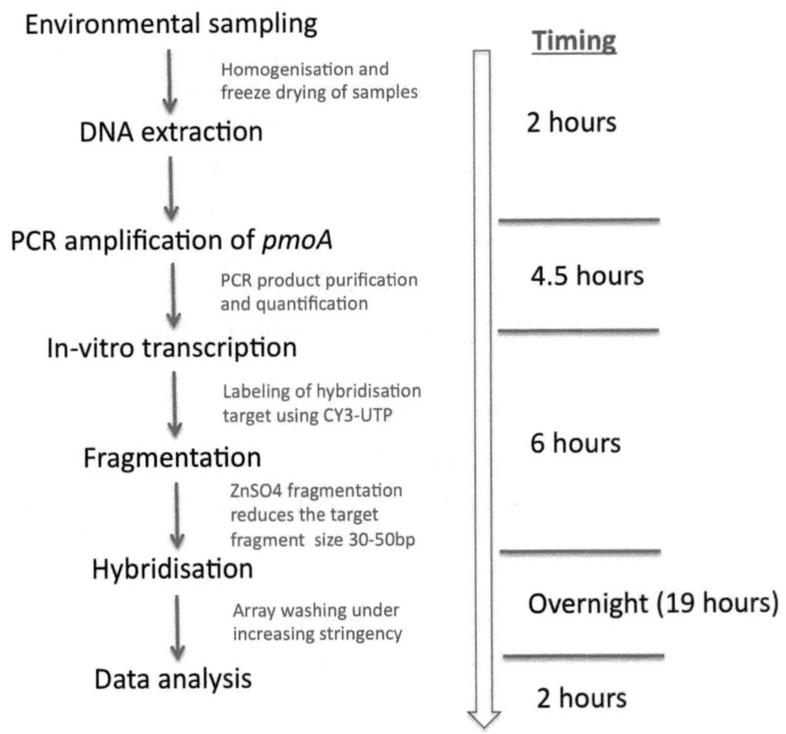

Fig. 1 Outline of the experimental procedure described in this chapter. Notes relating to each of the steps are included

3.1 Database Construction and Probe Design

1. A database of all publicly available *pmoA* sequences is created and used to create a new ARB database (www.arb-home.de) (*see* **Note 2**).
2. Sequences are aligned automatically and then corrected manually then used to create a PT server after which a neighbor-joining tree is created using 1,000 bootstraps.
3. Sequences are grouped into clades for which probes will be designed.
4. Probes are designed to individual as well as larger groups using the probe design function (*see* **Note 3**).
5. Probe specificity is checked in silico via CalcOligo (www.calcoligo.org). Probe _Match output files from ARB are fed into CalcOligo, resulting in an Excel sheet displaying the weighted mismatches for each probe versus each sequence in the database.

3.2 Array Printing

1. Probes are ordered to include a 5′ C6 amino modification and are subsequently eluted in sterile dH_2O.
2. A spotting plate comprising 50 nM probe in 50 % DMSO is set up for spotting.
3. Arrays are printed in triplicate (Fig. 2) on aldehyde slides at room temperature and 55 % humidity (*see* **Notes 4** and **5**).
4. Following printing, slides are dried in a dehumidification chamber (<30 % humidity) for 12 h prior to processing.

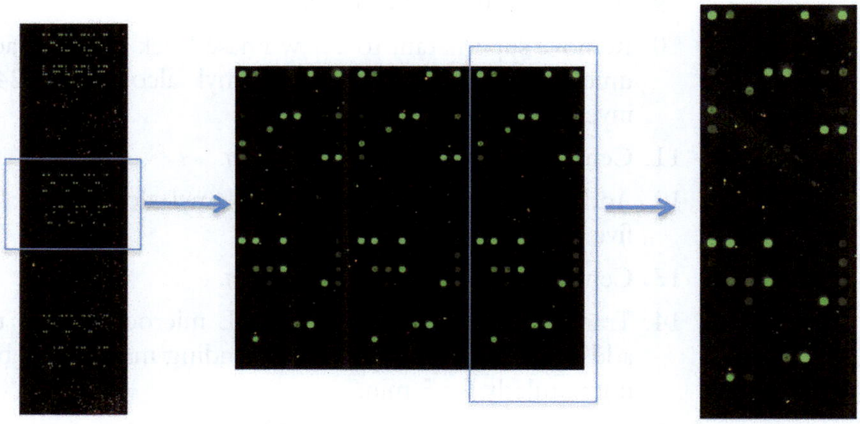

Fig. 2 Microarray slide layout, showing a typical image resulting from the hybridization of an environmental sample on the array. Three array sets are printed on each slide, allowing analysis of three samples per slide. Each of these array sets comprises the same array printed in triplicate

5. For processing, slides are washed with agitation twice in 0.2 % SDS for 2 min at room temperature, twice in dH_2O for 2 min at room temperature, and then soaked for 2 min in dH_2O at 95–100 °C to denature the DNA.
6. Slides are then allowed to cool at room temperature for 5 min after which they are treated with sodium borohydrate solution for 5 min.
7. Slides are rinsed three times with 0.2 % SDS for 1 min followed by one rinse in dH_2O for 1 min at room temperature.
8. Slides are dried by centrifugation at 500 × *g* and stored in the dark for up to 1 year (*see* **Notes 6** and **7**). *Stop point.*

3.3 DNA Extraction

1. Soil or sediment samples are collected and homogenized after which ~0.5 g is placed in a FastPrep tube (MP Biomedicals) and freeze-dried (*see* **Note 8**).
2. Add 780 μL of lysis buffer and incubate at 37 °C for 30 min.
3. Add 122 μL MT Buffer and agitate in a bead mill for 30 s at ~5.5 m/s.
4. Centrifuge for 2 min at 10,000 × *g*.
5. Collect supernatant and place into a (pre-spun) Phase Lock tube.
6. Add 500 μL of lysis buffer and 50 μL of MT Buffer to the FastPrep tube and agitate again at ~5.5 m/s for 30 s.
7. Centrifuge for 2 min at 10,000 × *g*.
8. Add supernatant to Phase Lock tube along with 5 μL of 10 mg/mL proteinase K.
9. Incubate at 65 °C for 30 min.
10. Remove supernatant to a new Phase Lock tube and add 1 volume of Phenol:chloroform:isoamyl alcohol (24:24:1) and invert five times.
11. Centrifuge for 2 min at 12,000 × *g*.
12. Add 1 volume of chloroform:isolamyl alcohol (24:1); invert five times.
13. Centrifuge for 2 min at 12,000 × *g*.
14. Transfer supernatant to a 1.7 mL microcentrifuge tube and add 1 volume of homogenized binding matrix; mix by inverting regularly for 5 min.
15. Centrifuge for 1 min at 10,000 × *g*, discard supernatant.
16. Resuspend supernatant in 500 μL of wash buffer and transfer to a spin filter.
17. Centrifuge the spin filter for 1 min at 10,000 × *g*; discard the eluate from the tube and add 500 μL of wash buffer.

18. Centrifuge for 1 min at 10,000 × *g*; discard the eluate and centrifuge again for 1 min to dry the tube.
19. Place the filter in a new 1.7 mL microcentrifuge tube and add 50 μL of elution buffer to the tube; incubate at room temperature for 1 min.
20. Centrifuge for 3 min at 12,000 × *g* to collect the DNA; discard the spin filter.
21. Following extraction, DNA can be quantified using a NanoDrop spectrophotometer (Peqlab, Erlangen, Germany) and subsequently stored at −20 °C for up to 2 years. *Stop point.*

3.4 PCR and IVT

1. *PCR amplification*: for each sample to be analyzed, set up three 25 μL PCR reactions containing the following (*see* **Note 9**): 2.5 μL of 10× PCR Buffer, 2 μL of 10 mM dNTPs, 1.5 μL of 50 mM $MgCl_2$, 1 μL of 1.5 μM 189F primer, 1 μL of 1.5 μM 682R primer, 0.2 μL of 5 U/μL Taq polymerase, 30 ng environmental DNA, and sterile dH_2O to a total volume of 25 μL.
2. Amplify the *pmoA* gene using the following thermal cycling conditions: an initial denaturation of 95 °C for 5 min followed by 10 cycles of 95 °C for 1 min, 65 (−1 °C per cycle) for 1 min and 72 °C for 1 min, then a subsequent 20 cycles of 95 °C for 1 min, 55 °C for 1 min, 72 °C for 1 min, and then a final cycle of 72 °C for 10 min.
3. Run 5 μL of each PCR product on a 1.5 % agarose gel to ensure successful amplification and fragment size (~500 bp).
4. Set up a second 25 μL PCR reaction containing the following: 2.5 μL of 10× PCR Buffer, 2 μL of 10 mM dNTPs, 1.5 μL of 50 mM $MgCl_2$, 1 μL of 1.5 μM 189F primer, 1 μL of 1.5 μM 661R-T7 primer, 0.2 μL of 5 U/μL Taq polymerase, and either 5 μL of 1/100 diluted PCR reaction from **step 2** (if a band was seen on agarose during **step 3**) or 5 μL of undiluted PCR product from **step 2** (if a band was not seen on agarose during **step 3**).
5. Repeat amplification of the *pmoA* gene using the following thermal cycling conditions: an initial denaturation of 95 °C for 5 min followed by 14 cycles of 95 °C for 1 min, 55 °C for 1 min, and 72 °C for 1 min, then a final cycle of 72 °C for 10 min (*see* **Note 10**).
6. Run 5 μL of each PCR product on a 1.5 % agarose gel to ensure successful amplification and fragment size (~500 bp).
7. Combine replicate PCR reactions from **step 5** and purify using the High Pure PCR Purification Kit, as per manufacturer's instructions; elute in 50 μL sterile of dH_2O.
8. Quantify the purified PCR product using the NanoDrop spectrophotometer. *Stop point.*

9. *In vitro transcription*: To a sterile 200 μL PCR tube, add 350 ng of purified PCR product from **step 7**, along with 1 μL of 50 ng/μL hyaB PCR product (*see* **Note 11**), 4 μL 5× T7 RNA polymerase buffer, 2 μL 100 mM DTT, 0.5 μL 40 U/μL RNasin, 1 μL of each of 10 mM ATP, CTP, and GTP, 0.5 μL of 10 mM UTP, 1 μL of 5 mM Cy3-UTP, 1 μL of 40 U/μL T7 RNA polymerase, and sterile dH_2O to a total volume of 20 μL.
10. Vortex the reaction to mix, then centrifuge briefly at 500 × *g*.
11. Incubate the reaction at 37 °C for 4 h.
12. Transfer the IVT reaction to a 1.7 mL microcentrifuge tube and add 80 μL of RNAse-free water followed by 350 μL of RLT buffer and 250 μL of 99 % ethanol; mix thoroughly.
13. Transfer solution to a RNeasy mini column and centrifuge at 1,000 × *g* for 30 s.
14. Transfer the column to a new collection tube and add 500 μL of RPE buffer; centrifuge for 30 s at 10,000 × *g*.
15. Transfer column to a new 1.7 mL microcentrifuge tube and add 50 μL of RNAse-free water; centrifuge for 1 min at 10,000 × *g*.
16. Transfer purified RNA to a new RNAse-free 1.7 μL microcentrifuge tube and add 1.43 μL of 1 M Tris–HCl and 5.71 μL of 100 mM $ZnSO_4$ and mix.
17. Incubate for 30 min at 60 °C (*see* **Note 12**).
18. Immediately place on ice and add 1.43 μL of 500 mM EDTA and 1 μL of RNasin (*see* **Note 13**). *Stop point.*

3.5 Hybridization

1. In an RNAse-free 1.7 mL microcentrifuge tube, add 63.5 μL of RNAse-free water, 1.11 μL of 10 % SDS, 4.42 μL of 50× Denhardt's reagent, 33.45 μL of 20× SSC, and 5 μL of fragmented IVT and mix.
2. Pre-warm array slides with hybridization chambers attached on a heating block at 55 °C.
3. Add hybridization solution to each of the HybriWells, seal the HybriWells, and then transfer the slides to a rack in preheated hybridization oven (55 °C) (*see* **Note 14**). It is best to process two slides at a time to ensure no drop in temperature (*see* **Note 15**).
4. Incubate slides overnight with slow rotation (~10 rpm) (*see* **Note 16**).
5. Following hybridization, remove the HybriWells from the slides one at a time and immediately place in 2× SSC and 0.1 % SDS and wash for 5 min with agitation (*see* **Note 17**).
6. Wash the slides twice for 5 min in 0.2× SSC, with shaking.
7. Wash the slides for 5 min in 0.1× SSC, with shaking.
8. Dry the slides with compressed air and store in the dark.

9. Scan slides within a few hours of processing using a GenePix 4000B scanner. Use maximum laser power and adjust PMT settings in a way that no saturated pixel appears on the array (*see* **Note 16**).

3.6 Data Analysis

1. Analyze scanned images first using GenePix software.
2. Paste the GenePix analysis output into Microsoft Excel, where further data analysis is carried out, using custom made or recorded macros.
3. Calculate the average "median minus background" intensities for each probe, over the three replicate spots on the array.
4. Normalize these results to that of the positive control (mtrof173), targeting the forward PCR primer pmoA189.
5. As quality indicators, consider percentage of saturated pixels (F532 % Sat.) and percentage of pixels above background + 2× standard deviation (%>B532 + 2SD).
6. Merge results from individual hybridizations in a summary Excel table, showing probes in columns and results for individual samples in subsequent rows.
7. Using the conditional formatting function of Excel, create a rough heatmap, indicating signal intensities (*see* **Note 18**). Consider maximum hybridization signals obtained during validation with pure targets as reference values for individual probes (reflecting their maximal hybridization capacities) and relate results to them when the heatmap is created (*see* **Notes 17** and **19**) (Fig. 3). For further information on statistical analysis, (*see* **Note 20**).

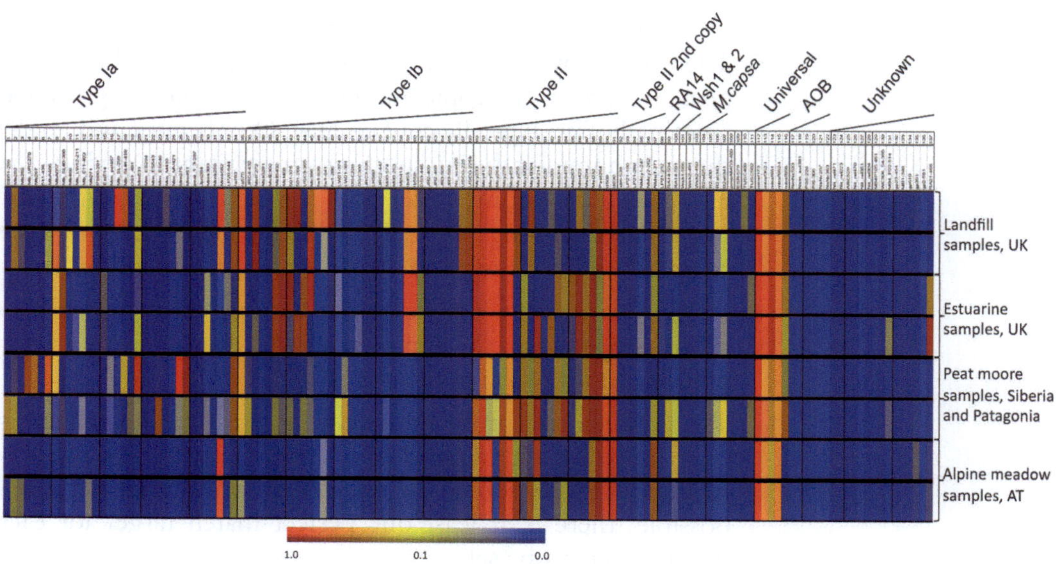

Fig. 3 Results of *pmoA* microarray analysis of samples from a number of different environments, including, landfill, estuarine sediment, peat moor soil, and alpine meadow soil. The probes are divided into different target groups (shown across the *top*)

4 Notes

1. The described methodology is also applicable to assaying the expression of *pmoA* genes, indicating not merely the presence but also the activity of the detected bacteria [10].
2. The ARB phylogenetic software package is used for creating and maintaining the sequence database, designing the probes and checking their predicted specificity in silico. Detailed instructions and information on the ARB software package is available from multiple sources, i.e., www.arb-home.de or www.arb-silva.de. A comprehensive sequence database and a robust phylogenetic tree are crucial for designing a quality probe set.
3. Optimal parameter settings for the probe design function will vary between databases and clades within databases. A good starting point, however, is shown below (only differences from default settings in ARB are shown):

   Max. nongroup hits—set to 5–10 % of the total number within the clade targeted.

   Min. group hits (%)—start with 70 %; drop if needed; can go down to as low to 10 %.

   Length—20–24 nt.

   Temperature—30–70.

   G + C content—35–70 %.
4. It is critical to maintain consistent and homogenous humidity within the microarray printer during spotting. Humidity settings are optimized for the slide surface and spotting buffer used. Changing the spotting buffer or the microarray surface may require re-optimization of humidity.
5. There are many hybridization buffers used; they can be roughly grouped into hygroscopic and non-hygroscopic. Hygroscopic spotting buffers typically contain DMSO or betaine and, under the right humidity during spotting, are resistant to drying. The 50 % DMSO spotting buffer used is a hygroscopic buffer and can be used in long (typically 24 h or longer) spotting runs repeatedly without need to dry down and resuspend probes.
6. Printed microarrays can be stored desiccated, in the dark, at room temperature, for approximately 6 months without significant deterioration in quality.
7. The final, spotted probe set is subjected to rigorous validation by hybridizing with a panel of reference targets. The panel of reference targets should be selected in a way that whenever possible, there is at least one perfect match target for each probe in the probe set.
8. The DNA extraction method applied has a profound effect on results. In order to be able to directly compare results, a

common DNA extraction protocol has to be established and followed rigorously. For details *see* Pan et al. [9].

9. It is advisable to run three parallel PCR experiments rather than a single larger batch in order to minimize the potential for random drifts in the PCR influencing results.
10. PCR approach: The majority of samples analyzed using the *pmoA* array employ the semi-nested PCR approach described here. It is also possible to proceed with target preparation and hybridization using PCR product from the first round only (**steps 1** and **2**). This usually results in a better coverage of ammonia oxidizing bacteria (*amoA*) as this primer set will amplify both *pmoA* and *amoA*.
11. The *hyaB* PCR product is an external spike used for control purposes. The amplicon is of a hydrogenase gene from *E. coli*. Amplification conditions are 95 °C for 5 min; 35 cycles of 95 °C for 1 min, 56 °C for 1 min, and 72 °C for 1 min; and a final cycle of 72 °C for 10 min.

    Primers: hyaB1 (GACCCGATTACGCGCATCGAAGG) and T7-hyaB2 (TAATACGACTCACTATAGCCAGTAGCCATT GCGGAAGATCC).
12. During the fragmentation as described in the protocol, condensation forms on the lid of the Eppendorf tubes. This is part of the protocol, and preventing it from happening (i.e., incubating tubes in a water bath or in a PCR machine with heated lid) may influence the fragmentation efficiency, likewise changing the volume of the Eppendorf tubes.
13. Fragmented RNA targets supplemented with RNAsin can be stored at −20 °C for at least 2 years without significant deterioration in quality.
14. When adding the hybridization mixture, pay attention not to introduce air bubbles into the HybriWell. This can be aided by keeping the slides in a slightly tilted position, where the hybridization mixture is added through the opening in the lower position, filling the chamber towards the other opening in the highest position.
15. Check the temperature of the hybridization oven with a thermometer and make sure it is consistent throughout the hybridization procedure.
16. Cy3 (as all the fluorescent dyes on the market) is light sensitive. Carry out hybridization in the dark. Scanning causes a small amount of bleaching of the spots. Never stop scanning halfway through an array as this may introduce artifact.
17. It is crucial not to allow the hybridization mixture dry onto the slide. Cy3 adheres extremely well to glass if allowed to dry. When removing HybriWells, dip slides immediately into the first solution.

18. For publication quality figures, a proper heatmap can be created using GeneSpring software (Agilent, Santa Clara, CA, USA), following the same principles as used for the Excel heatmap.
19. Mismatches are weighted according to their position and nature. These weights are determined empirically and may differ between microarray methods used (RNA or DNA target, length of probes, length of spacers, surface property of the array slide, etc.). The parameters we use are Positions: 5′ first 0.3; 5′ second 0.6; 5′ third 1.0; 3′ first 0.3; 3′ second 0.8; 3′ third 1.1; all other positions 1.2. Mismatched base pairs: dArC 1.2; dTrC 1.2; dGrU 0.7; dTrG 0.4; all other mismatched base pairs 1.0. (Probes are DNA, targets RNA.)
20. Multivariate statistical analysis of array data allows the interpretation of results from a large number of samples. Approaches that combine the array data with other (abiotic) datasets are well suited to elucidating the drivers of methanotroph diversity. A good summary of the different analysis methods, including their advantages and limitations is given by Ramette [11].

Acknowledgement

This research was supported by the ESF EuroDiversity programme METHECO (No. FP018, local funding agencies: FWF, Austria, project number I40-B06). We would also like to thank Dr. Deepak Kumaresan for critical evaluation of the manuscript.

References

1. Bodrossy L et al (2003) Development and validation of a diagnostic microbial microarray for methanotrophs. Environ Microbiol 5:566–582
2. Stralis-Pavese N et al (2004) Optimization of diagnostic microarray for application in analysing landfill methanotroph communities under different plant covers. Environ Microbiol 6:347–363
3. Stralis-Pavese N et al (2011) Analysis of methanotroph community composition using a pmoA-based microbial diagnostic microarray. Nat Protoc 6:609–624
4. Abell GCJ et al (2009) Grazing affects methanotroph activity and diversity in an alpine meadow soil. Environ Microbiol Rep 1: 457–465
5. Han B et al (2009) Diversity and activity of methanotrophs in alkaline soil from a Chinese coal mine. FEMS Microbiol Ecol 70:40–51
6. Moussard H et al (2009) Identification of active methylotrophic bacteria inhabiting surface sediment of a marine estuary. Environ Microbiol Rep 1:424–433
7. Chen Y et al (2008) Diversity of the active methanotrophic community in acidic peatlands as assessed by mRNA and SIP-PLFA analyses. Environ Microbiol 10:446–459
8. Kip N et al (2010) Global prevalence of methane oxidation by symbiotic bacteria in peat-moss ecosystems. Nat Geosci 3:617–621
9. Pan Y et al (2010) Impacts of inter- and intralaboratory variations on the reproducibility of microbial community analyses. Appl Environ Microbiol 76:7451–7458
10. Bodrossy L et al (2006) mRNA-based parallel detection of active methanotroph populations using a diagnostic microarray. Appl Environ Microbiol 72:1672–1676
11. Ramette A (2007) Multivariate analyses in microbial ecology. FEMS Microbiol Ecol 62: 142–160

Chapter 10

Biolog Phenotype MicroArrays for Phenotypic Characterization of Microbial Cells

Amanda M. Mackie, Karl A. Hassan, Ian T. Paulsen, and Sasha G. Tetu

Abstract

Biolog Phenotype MicroArrays for microorganisms provide a high-throughput method for the global analysis of microbial growth phenotypes. Using a colorimetric reaction that is indicative of respiration, these microplate assays measure the response of an individual strain or microbial community to a large and diverse range of nutrients and chemicals. Phenotype MicroArrays have been used to study gene function and to improve genome annotation in single microorganisms and for physiological profiling of bacterial communities. The microplate system can be used to obtain a comprehensive overview of metabolic capability, or it can be tailored, through the use of subsets of plates, to address specific research needs.

Key words Phenotype, Biolog, Metabolic fingerprinting, Microplate assays, Susceptibility testing

1 Introduction

Biolog Phenotype MicroArrays (PMs) are commercially available microplate assays that can be used to test more than 1,000 phenotypic traits simultaneously by recording an organism's respiration over time on many distinct substrates [1, 2]. Although designed for pure cultures, these types of assays can also be tailored to the study of microbial communities. Analysis using the full set of 20 microarray plates provides a comprehensive set of growth phenotypes including information on the ability to metabolize approximately 200 carbon sources, 400 nitrogen sources, and 100 phosphorous and sulfur sources plus sensitivity to 240 individual drugs and chemicals and to variations in osmolarity and pH.

Assaying phenotypes using the Biolog system involves adding a cell suspension to the 96 wells of a PM microplate, each of which contains the necessary ingredients to create unique culture conditions. For example, the 190 test wells of PM plates 1 and 2A each contain a single alternative carbon source along with all the other necessary ingredients for cell growth that are not included in the media. If the organism is able to transport and catabolize the

Ian T. Paulsen and Andrew J. Holmes (eds.), *Environmental Microbiology: Methods and Protocols*, Methods in Molecular Biology, vol. 1096, DOI 10.1007/978-1-62703-712-9_10, © Springer Science+Business Media, LLC 2014

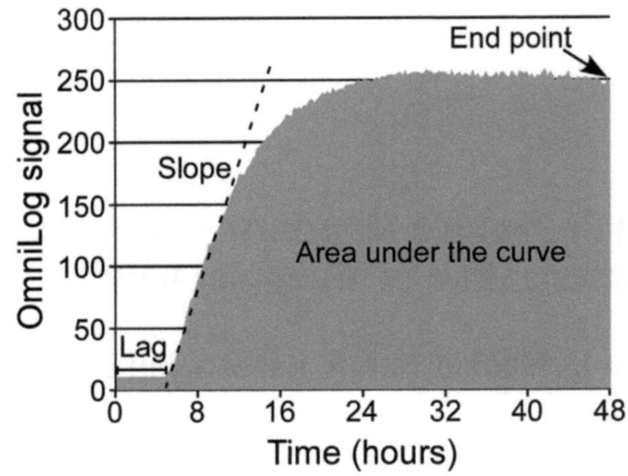

Fig. 1 Parameters describing cell growth. *E. coli* DH5α cells were incubated in IF-10a (+Dye A) at 37 °C in Biolog plate PM19 for 48 h. Well intensity data were collected every 15 min using an OmniLog plate reader, and the results from well C10 (cinnamic acid) are shown. Parameters including the initial lag phase (Lag), the initial slope of the curve (Slope), the area under the curve, and the final well intensity (End point) can be used to describe the growth characteristics for comparison to other strains or conditions

particular carbon source provided, respiration will occur. The Biolog assays use a tetrazolium redox dye to measure NADH formation as a sensitive indicator of respiration—reduction of the dye results in the formation of purple formazan products in the well, and this colorimetric reaction can be monitored and recorded using specialized instrumentation provided by Biolog (the OmniLog PM system, which records the color change every 15 min for up to 50 plates). Using this, a kinetic response curve which parallels microbial growth can be generated for each well, allowing growth to be compared between samples through multiple parameters such as lag, slope, and area under the curve (Fig. 1). Reactions can also be monitored using a microplate reader [3, 4], or for some applications, a simple visual assessment of color formation at the end point of incubation may suffice.

Biolog PMs have found application in a variety of fields including microbial characterization, ecology, and systems biology [5]. One common application of the PM system is to detect phenotypic changes associated with gene knockouts, and this strategy has been used to study gene function and to assess and improve genome annotation [6, 7]. In our laboratory, Biolog PMs enabled the successful phenotypic characterization of 27 *Pseudomonas aeruginosa* knockout mutants, defective in sugar or amino acid transport, in a single study [7], emphasizing the relatively high-throughput nature of this approach relative to traditional characterization methods. We have also used this approach to assess the carbon utilization profile of *E. coli* transporter knockouts (Fig. 2) and to characterize

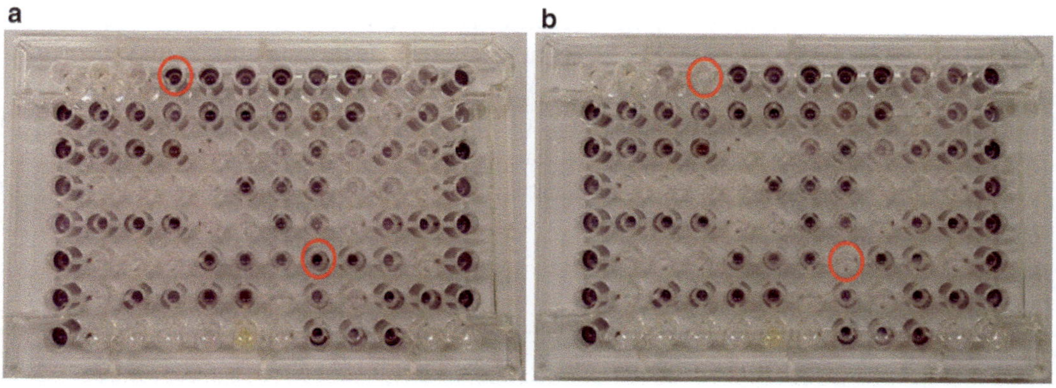

Fig. 2 An example of PM plates (PM1 and 2A) following color development assays. Carbon utilization phenotypes of *E. coli* K-12 (**a**) compared to a double knockout mutant containing deletions of the *gudP* and *garP* transport genes (**b**). Circles indicate wells containing D-saccharic acid (row 1, column 4) and mucic acid (row 6, column 8) as sole carbon source. Lack of purple color in these wells in B is due to deletion of the *gudP* and *garP* genes encoding putative glucarate transporters in *E. coli* K-12. The double deletion strain is unable to transport the substrate; there is no cellular respiration, and the indicator dye is not reduced to a purple formazan product. The yellow color observed in some wells is due to the specific substrate present in that well

novel drug efflux systems. PMs can be used to help understand the metabolic capabilities and stress susceptibilities of different bacteria. A collection of related microbial strains can be readily assayed to identify differences in nutrient utilization or chemical sensitivity, as carried out recently for a number of newly genome sequenced plant-associated *Pseudomonas* species [8]. PM plates have also been used in less conventional ways. A recent publication, for example, describes adapting the PM plate system to assay differences in *E. coli* biofilm formation on a range of carbon sources [9].

Biolog PMs have been successfully applied to microbial community-level physiological profiling to study spatial and temporal differences in microbial communities from a wide range of environments including soil, water, wastewater, and industrial waste. The use of Biolog microplate assays for community analysis, via direct incubation of whole environmental samples rather than pure cultures, was first described by Garland and Mills [10]. With this approach, environmental samples are inoculated directly into plates either as aqueous samples or following suspension (care must be taken to standardize initial inoculum densities, which have been reported to be as low as ~10^4 cfu/mL [11]). The temperature chosen for plate incubation, frequency of data collection, and total running time of experiment will depend on the environment being sampled. For such applications, a subset of PM plates is generally used, most commonly carbon source utilization plates PM1 and 2A. Alternatively the Biolog EcoPlate, containing a set of only 31 carbon sources repeated three times on a single plate, can be used to generate a simple "metabolic fingerprint" or custom plates containing specific compounds of interest may be generated using

Biolog MT microplates (*see* **Note 1**). The results of these community profiling assays are generally analyzed using principal component analysis (PCA) of average well color development data, although other methods have also been used [12–14]. After color development data has been collected, it is also possible to gain insight into the microbial population contributing to different metabolic profiles by using cell lysate from wells of interest as template for PCR-DGGE analysis [3, 15].

2 Materials

2.1 Specialized Materials Purchased from Biolog

1. Biolog PM microplates (PM1–20).
2. IF-0a GN/GP base inoculating fluid (1.2×).
3. IF-10a GN or IF-10b GN/GP base inoculating fluid (1.2×) (*see* **Note 2**).
4. Biolog redox dye mix (100×) (*see* **Note 3**).
5. OmniLog microplate incubator/reader or variable temperature incubator and microplate reader.

2.2 General Materials

1. Agar plates for pre-growing microorganism of interest (*see* **Note 4**).
2. Sterile cotton swabs.
3. Sterile 30–50 and 150 mL containers.
4. Multichannel pipette and sterile 1.25 mL sterile pipette tips.
5. Sterile 50 mL reservoirs (if using a multichannel pipette).
6. Sterile capped test tubes—for preparing initial cell suspensions and measuring turbidity.
7. Turbidimeter or spectrophotometer.
8. 2 M carbon source stock solution (filter sterilized) for use with PM plates 3B–5 (and PM plates 6–8 if required).

3 Methods

The basic protocol for phenotype testing using Biolog PM plates involves preparing a standard suspension of bacteria in an inoculating fluid containing a tetrazolium dye and then adding a fixed amount to individual wells of a microplate. In our laboratory, we have found that it is unusual for a researcher to undertake phenotype testing using the full complement of 20 PM microplates at the one time, and thus we have organized this section to describe methods for using three different subsets of Biolog PM microplates, namely, PM plates 1 and 2A, PM plates 3B–5, and PM plates 9–20 (*see* also Table 1 which includes the assay set up for several additional plate combinations). Once a researcher is familiar

Table 1
Assay setup for varying combinations of Biolog PM microplates

	PM 1 and 2	PM 3–5	PM 3–8	PM 9 and 10	PM 11–20	PM 9–20
IF-0a (1.2×)	15.25	22.25	44.5	–	–	–
IF-10 (1.2×)	–	–	–	18.3	83.3	100
Dye mix (100×)	0.22	0.32	0.64	0.22	1	1.2
H_2O	2.83	3.81	7.62	3.37	15.2	18.2
2 M carbon source stock solution	–	0.32	0.64	–	–	–
42 % T cell suspension	3.7	5.3	10.6	–	–	–
85 % T cell suspension	–	–		0.11	0.5	0.6
Final volume	22	32	64	22	100	120

All volumes are in mL. Final volume amounts are sufficient to inoculate the number of plates in the given series using 100 μL per well plus excess

with the use of the Biolog testing procedure, it is easy to adapt the protocol to test any combination of plates that may be relevant to particular needs.

3.1 Phenotype Testing Using Plates PM1 and 2A (Carbon Utilization Plates)

1. Streak the bacteria on appropriate agar plates (e.g., R2A, LB, sheep blood) and grow overnight at a suitable temperature. A second subculturing is recommended if the cells have been removed from a frozen stock.
2. Pipette 12.5 mL of Biolog IF-0a media (1.2× concentration) into a sterile capped test tube and add 2.5 mL water. Remove 10 mL of this solution (now at 1× concentration) to a second sterile capped test tube. The remaining 5 mL should be put to one side and can be used, if necessary, for sample dilution when adjusting the density of the cell suspension in **step 3**.
3. Using a sterile swab, remove several colonies from the agar plate prepared in **step 1** and transfer into the test tube containing 10 mL IF-0a (*see* **Note 5**). Mix the suspension gently but thoroughly. Check the turbidity in the Biolog turbidimeter and adjust (by adding either more cells or more 1× IF-0a) to achieve 42 % transmittance (*see* **Note 6**). Place the cell suspension to one side while preparing the "IF-0a plus dye" mix.
4. For the IF-0a plus dye mix, remove 15.25 mL Biolog IF-0a media (1.2× concentration) to a sterile 30–50 mL container and add 0.22 mL Biolog dye mix (100×) and 2.83 mL water.
5. Prepare a 1:5 dilution of cells by adding 3.7 mL of the cell suspension prepared in **step 3** to the IF-0a plus dye mix. Mix gently but thoroughly. This final cell suspension will have 85 % transmittance or an absorbance of 0.07 at 600 nm.

6. Inoculate PM plates 1 and 2A with 100 µL of the final cell suspension per well. If you are using a multichannel pipette, transfer the cell suspension to a sterile reservoir before use.
7. Incubate the PM plates for 24–48 h using the same temperature at which the organism was grown. If you are using the OmniLog incubator, readings of each well are taken every 15 min for the specified time period. The associated software can then be used to view and edit data, to compare data lists, and to generate reports. Typically, reports generated using the OmniLog software will display data as kinetic plots which represent the change in respiration/growth over time for each individual well. Depending on the experimental design, kinetic plots can also be overlaid, thus providing visual cues to phenotype variation. Open-source software for the analysis of kinetic data generated by the OmniLog software has also recently been reported [16].

3.2 Phenotype Testing Using Plates 3B–5 (Nitrogen, Phosphorous, Sulfur, and Other Nutrient Utilization)

PM plates 3B and 4A are designed to test the nitrogen, phosphorous, and sulfur utilization capabilities of a microorganism, while PM plate 5 assays utilization of various nutrient supplements. The method for phenotype testing using these plates is similar to that used for PM plates 1 and 2A; however, a carbon source must be supplied as part of the inoculation media when using PM plates 3B–5. The protocol supplied by Biolog includes the use of succinate or pyruvate as a carbon source (which is effective for *E. coli*); however, this may vary depending on the particular microorganism being studied. In practice, the results obtained from carbon utilization testing using PM plates 1 and 2B may give a clearer indication of the preferred carbon source(s) for a test organism.

1. Prepare a cell suspension in the same manner as outlined for PM plates 1 and 2A (**steps 1–3** of Subheading 3.1).
2. Prepare an "IF-0a plus dye" mix suitable for use with PM plates 3B–5A. Remove 22.25 mL Biolog IF-0a media (1.2× concentration) to a sterile 50 mL container, and add 0.32 mL Biolog dye mix (100×), 0.32 mL of the 2 M preferred carbon source stock solution, and 3.81 mL water.
3. Prepare a 1:5 dilution of cells by adding 5.3 mL of the cell suspension prepared in **step 1** to the IF-0a plus dye mix. Mix gently but thoroughly. This final cell suspension will have 85 % transmittance or an absorbance of 0.07 at 600 nm.
4. Inoculate PM plates 3B, 4A, and 5 with 100 µL cell suspension per well. Incubate and analyze as described for PM plates 1 and 2A above.

3.3 Phenotype Testing Using Plates 9–20 (Chemical and Drug Sensitivity)

PM plates 9 and 10 are designed to assay sensitivity to variations in osmolarity and pH, while PM plates 11–20 contain 240 different chemicals each at four differing concentrations. The method below is for use with the full 12 plates but could be modified to suit the use of a particular subset of these plates (e.g., plates 9 and 10 only) if desired (*see* Table 1).

1. Prepare a cell suspension in the same manner as outlined for PM plates 1 and 2A (**steps 1–3** of Subheading 3.1) except with a final transmittance of 85 % or an absorbance of 0.07 (*see* **Note** 7).
2. Prepare an "IF-10 plus dye" mix by removing 100 mL Biolog IF-10 (1.2× concentration) to a sterile 150 mL container and add 1.2 mL Biolog dye mix (100× concentration) and 18.8 mL water.
3. Prepare a 1:200 dilution of cells by adding 0.6 mL of the cell suspension prepared in **step 1** to the IF-10 plus dye mix. Mix gently but thoroughly.
4. Inoculate PM plates 9–20 with 100 μL cell suspension per well. Incubate and analyze as described previously.

4 Notes

1. Custom plates can be designed to assay phenotypes of particular interest. Biolog MT2 MicroPlates are suitable for metabolic testing (e.g., carbon utilization profiling) in combination with specific inoculation fluids. These plates contain only a nutrient base and redox dye, so putative substrates can be added as desired. Alternatively, we have developed bacterial susceptibility tests using regular commercial 96-well microplates. Antimicrobial compounds of interest were added to the plates at varying concentrations. Cells were added to the plates suspended in Biolog IF-10a plus dye and monitored following the standard protocols set out in Subheading 3.3.
2. IF-0a GN/GP base is recommended for use with both Gram-negative and Gram-positive bacteria. IF-10a GN base is recommended for use with Gram-negative bacteria, and IF-10b GN/GP base is recommended for use with Gram-positive bacteria but is also suitable for use with Gram-negative organisms.
3. Biolog redox dye mixes are proprietary tetrazolium-based dyes, the reduction of which results in the production of a purple formazan product. The use of tetrazolium dyes as indicators of cell metabolism has been comprehensively reviewed [17]. Biolog supplies eight redox dyes (A–F) and recommendations for use.

4. Prior to setting up assays, bacteria are pre-grown overnight on solid media. Biolog recommends a universal growth agar plus blood for organisms requiring high nutrient amounts and R2A agar for those able to grow with less nutrients. LB agar is also suitable for use. We have observed variations in assay results due to differing pre-growth conditions.
5. Pre-wetting the swab with sterile water allows cells to be collected easily from the solid media with minimal agar transfer.
6. A cell suspension can be prepared based on absorbance (at 600 nm) if a turbidimeter is not available using the equation: $A = 2 - \log_{10}\%T$.
7. We recommend preparing a cell suspension with an absorbance of 0.37 and then diluting 1:5 to achieve the final desired density of 0.07 absorbance or 85 % transmittance. IF-0a at 1× concentration should be used for preparing the cell suspension.

References

1. Bochner BR, Gaddzinski P, Panomitros E (2001) Phenotype MicroArrays for high-throughput phenotypic testing and assay of gene function. Genome Res 11:1246–1255
2. Bochner BR (2003) New technologies to assess genotype-phenotype relationships. Nat Rev Genet 4:309–314
3. İnceoğlu Ö, Falcão Salles J, van Elsas DJ (2011) Soil and cultivar type shape the bacterial community in the potato rhizosphere. Microb Ecol 63:460–470
4. Barua D, Kim J, Reed JL (2010) An automated phenotype-driven approach (GeneForce) for refining metabolic and regulatory models. PLoS Comput Biol 6:e1000970
5. Bochner BR (2009) Global phenotypic characterization of bacteria. FEMS Microbiol Rev 33:191–205
6. Loh KD et al (2006) A previously undescribed pathway for pyrimidine catabolism. Proc Natl Acad Sci U S A 103:5114–5119
7. Johnson DA et al (2008) High throughput phenotypic characterisation of *Pseudomonas aeruginosa* membrane transport genes. PLoS Genet 4:1–11
8. Loper JE et al (2012) Comparative genomics of plant-associated *pseudomonas* spp.: insights into diversity and inheritance of traits involved in multitrophic interactions. PLoS Genet 8:e1002784
9. Pruss BM et al (2010) Environmental and genetic factors that contribute to Escherichia coli K-12 biofilm formation. Arch Microbiol 192:715–728
10. Garland JL, Mills AL (1991) Classification and characterization of heterotrophic microbial communities on the basis of patterns of community-level sole-carbon-source utilization. Appl Environ Microbiol 57:2351–2359
11. Insam H et al (1996) Changes in functional abilities of the microbial community during composting of manure. Microb Ecol 31: 77–87
12. Garland JL (1996) Analytical approaches to the characterization of samples of microbial communities using patterns of potential C source utilization. Soil Biol Biochem 28: 213–221
13. Hackett CA, Griffiths BS (1997) Statistical analysis of the time-course of Biolog substrate utilization. J Microbiol Meth 30:63–69
14. Harch BD et al (1997) Using the Gini coefficient with BIOLOG substrate utilisation data to provide an alternative quantitative measure for comparing bacterial soil communities. J Microbiol Meth 30:91–101
15. Smalla K et al (1998) Analysis of BIOLOG GN substrate utilization patterns by microbial communities. Appl Environ Microbiol 64: 1220–1225
16. Vaas LA et al (2012) Visualization and curve-parameter estimation strategies for efficient exploration of phenotype microarray kinetics. PLoS One 7:e34846
17. Berridge MV, Herst PM, Tan AS (2005) Tetrazolium dyes as tools in cell biology: new insights into their cellular reduction. Biotech Ann Rev 11:127–152

Part III

Investigation of Biological Activities In Situ

Chapter 11

Visualization of Metabolic Properties of Bacterial Cells Using Nanoscale Secondary Ion Mass Spectrometry (NanoSIMS)

Yi Vee Chew, Andrew J. Holmes, and John B. Cliff

Abstract

NanoSIMS combines high-resolution imaging and mass spectrometry with simultaneous collection of up to seven different masses, providing an invaluable technique for determining the isotopic and elemental composition in microscopic target samples. It has been used in varying fields, from studying the elemental composition of mineral samples to tracking cell uptake of isotope-labelled substrates. In combination with in situ hybridization techniques, NanoSIMS offers a powerful method of linking metabolic capacity to phylogenetic identity in cell samples. Here, we describe methods and considerations for microbial sample preparation, visualization, and analysis using NanoSIMS.

Key words NanoSIMS, Subcellular localization, Stable isotope enrichment, Halogenous molecules, Cell imaging, Isotope tracking, Elemental tracking, Fluorescence in situ hybridization (FISH)

1 Introduction

Secondary ion mass spectrometry (SIMS) is an ion microprobe technology based on mass spectrometry of secondary ions extracted from the surface of a solid sample under the impact of an energetic beam of primary ions, providing spatially resolved information on molecular and isotopic properties of the target [1].

SIMS can be divided into static or dynamic SIMS. With static SIMS, measurements are performed with a number of incident ions less than one order of magnitude of the number of atoms at the surface of the sample. Characterization of the sample surface may be accomplished through analysis of atomic or molecular fragments indicative of the sample. The resulting information is restricted to relatively abundant species in the superficial layers (<1 nm) [2]. With dynamic SIMS, the number of incident ions typically exceeds the number of surface atoms on the sample allowing isotopic/elemental information to be obtained at high count rates.

Ian T. Paulsen and Andrew J. Holmes (eds.), *Environmental Microbiology: Methods and Protocols*, Methods in Molecular Biology, vol. 1096, DOI 10.1007/978-1-62703-712-9_11, © Springer Science+Business Media, LLC 2014

By ablating a crater in the sample, depth-profiling data may be acquired to a depth of a few μm below the surface. A variety of primary beam/mass spectrometer configurations are available commercially. The NanoSIMS is typically used in dynamic mode and uses a normal incident primary ion beam and a double-focusing, electric-sector/magnetic-sector mass spectrometer and is optimized for high spatial resolution along with high transmission at high mass resolving power. Because picoampere primary currents are typically used for imaging, NanoSIMS depth profiles are typically limited much less than a micron.

Analysis occurs under ultrahigh vacuum (as low as to 10^{-10} Torr) to prevent atmospheric interference with primary and secondary ions and to prevent high-voltage arcing. Secondary ion and electron emission occurs when high-energy particles collide with the sample. The primary ion beam energy dissipates into the target, causing a collision cascade and releasing secondary ions from atomic layers near the sample surface. A series of lenses, deflectors, and stigmators extract these secondary ions and direct the secondary beam into the mass spectrometer. Discrete masses are directed into individual detectors which when imaging are electron multipliers. By correlating primary beam position with secondary signal, NanoSIMS allows us to build a 2D or even 3D map of the spatial distribution of isotopes in a sample (Fig. 1).

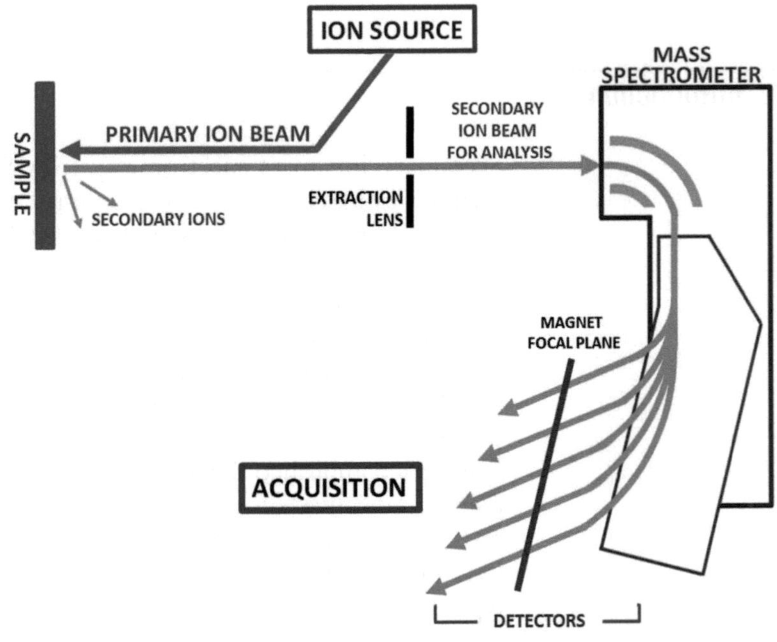

Fig. 1 Simplified schematic of a NanoSIMS instrument showing major components for biological imaging. Collision of a Cs^+ or O^- primary beam with the sample surface releases secondary ions and electrons which are then directed into a mass spectrometer for mass detection and image acquisition

Early SIMS ion probes were highly destructive to samples with low lateral resolution and mass separation power. Increases in mass or spatial resolution resulted in a trade-off with sensitivity. A major advance was the development of a fine focus primary ion beam, moved line by line across the sample surface for improved resolution [3, 4]. Collaboration between CAMECA (a company involved in development of SIMS instruments), University of Paris-Sud, Orsay, and the French Space Agency resulted in the CAMECA NanoSIMS 50, an ion microprobe with high mass and lateral resolution and sensitivity—suitable for biological applications [5]. The CAMECA NanoSIMS 50 allows high lateral resolution using either a Cs^+ (≥50 nm) or O^- (≥150 nm) primary beam. Up to five masses (or seven for the newer NanoSIMS 50 L) can be measured in parallel from the same sample layer, allowing superposition of ion data. The NanoSIMS 50 also allows indirect optical imaging of the sample to select areas of interest for analysis.

NanoSIMS has already been used for a variety of applications in physical and life sciences including the study of meteorites [6, 7], corrosion in alloys [8], nutrient uptake in the rhizosphere [9], uptake of isotope-labelled drugs by cancer cells [10], and tracking metabolic function and exchange in cells [11–13].

In environmental microbiology, NanoSIMS is proving particularly useful for linking microbial phylogeny to metabolism [14–17]. It allows the tracing of isotopically labelled macronutrients to an individual cell, or in the case of eukaryotic microorganisms, NanoSIMS allows the possibility of tracking labelled nutrients to subcellular components. NanoSIMS can also be coupled to in situ hybridization to permit identification of cells in mixtures. In this chapter, we outline general experimental design considerations for NanoSIMS in environmental microbiology and detail tracking of gut microbial uptake of isotope-labelled substrate supplied intravenously to the host.

2 Materials

Prepare all reagents using ultrapure water (water filtered and deionized to a resistivity of 18.2 MΩ cm at 25 °C). All reagents should be prepared and stored at room temperature unless indicated otherwise. Store fluorescent components in the dark and reduce light exposure to a minimum.

2.1 Biological Samples

1. Labelling substrate: (Isotope/element selection considerations are outlined in Subheading 3.1, **step 1**). In this experiment, we are using isotopically labelled threonine (98 % ^{13}C, 98 % ^{15}N).
2. Biological samples and controls: The cells must be obtained or cultured in the presence (test cells) and absence (control cells) of labelling substrate. Unrelated unlabelled control cells provide an

additional control for environmental levels of the isotopes in question. In this experiment, the biological sample is gut bacteria. The test samples are from the gut of host mice injected with isotopic threonine as a test sample, and the control samples are gut bacteria from uninjected mice. We also used wild-type laboratory-grown yeast cells as isotopic standards.

3. Incubation system: This can be any system of controlled supply of labelled substrate. For microbial cells, in vitro feeding experiments are readily done by batch culturing cells in the presence of labelled substrate, or in this case, an in situ experiment was done by injecting the host mice with labelled substrate and tracking uptake in gut bacteria. Additional experimental design considerations are outlined in Subheading 3.1, **step 2**.

2.2 Cell Fixation Components

1. Phosphate-buffered saline (PBS): 137 mM NaCl, 2.7 mM KCl, 4.5 mM Na_2HPO_4, and 1.47 mM KH_2PO_4. Dissolve 8 g NaCl, 0.2 g KCl, 1.44 g Na_2HPO_4, and 0.24 g KH_2PO_4 in 800 mL water. Adjust pH to 7.4, and then add water to 1 L final volume. Sterilize. Store at 4 °C.
2. 4 % (w/v) paraformaldehyde fixative (PFA): Dissolve 4 g PFA in 90 mL 1× PBS. Heat to 58 °C in the fume hood (*see* **Note 1**). Add 10 M NaOH to clear the solution (*see* **Note 2**). Remove from heat and adjust pH to 7.0–7.5. Add 1× PBS to 100 mL final volume. Filter sterilize through a 0.22 μm filter. Store at 4 °C (*see* **Note 3**).
3. 50 % (v/v) ethanol:PBS: Combine equivalent volumes of absolute ethanol and PBS. Store at 4 °C.

2.3 Fluorescence In Situ Hybridization (FISH) Components

NanoSIMS analysis can be combined with fluorescent in situ hybridization (FISH) to link metabolic capacity to phylogenic identity using fluor-labelled oligonucleotide probes. It is possible to do tandem analyses with microautoradiography (MAR)–FISH allowing observation of radioisotope incorporation into microbial cells in a community. It is also possible to use catalyzed reporter deposition (CARD)–FISH with horseradish peroxidase (HRP)-labelled probes and halogen-containing tyramides allowing boosting of a halogenic substrate signal for greater sensitivity in NanoSIMS analysis [14, 18]. In this example, we link microbial identity to substrate uptake using a group-specific phylogenetic probe:

1. Phylogenetic probes 3′-conjugated to a fluorophore (fluorescein and CY3 are commonly used) and halogen-labelled for NanoSIMS analysis at the 5′ end (*see* Subheading 3.1, **step 1**), e.g., Bacto1080 (Bacteroides group-specific probe) 5′-[5.5IdU] GCACTTAAGCCGACACCT[36-FAM]-3′: Dilute probe to a working concentration 50 ng/μl in water.

2. PBS (*see* Subheading 2.2, **item 1**).
3. 0.02 M ethylenediaminetetraacetic acid (EDTA): Add 0.74 g EDTA to 100 mL water. Sterilize.
4. Hybridization buffer: 0.9 M NaCl, 20 mM Tris–HCl, 0.1 % (w/v) SDS, and 20 % formamide. Make up a 1 M stock solution of Tris–HCl by combining 121.1 g Tris base with 60 mL HCl and pH to 7.5. Make up a 10 % (w/v) stock solution of SDS by combining 10 g SDS in 100 mL of water. Heat SDS solution to 68 °C to solubilize. Combine 52.6 g NaCl, 20 mL of stock 1 M Tris–HCl, and 10 mL of stock 10 % SDS in 750 mL of water. Add 250 mL formamide to make 1 L of hybridization buffer. Sterilize.
5. Wash buffer: 0.215 M NaCl, 20 mM Tris–HCl, 5 mM EDTA, and 0.1 % (w/v) SDS. Combine 12.5 g NaCl, 20 mL of stock 1 M Tris–HCl, 1.48 g EDTA, and 10 mL of stock 10 % SDS in 1 L of water. Sterilize.

2.4 Sample Mounting

Samples may be immobilized on various types of support material depending on the experimental aim, substrate used, and sample type. The most commonly used are silicon wafers or polycarbonate filters. NanoSIMS holders can take flat round discs of 10 mm, 13 mm, and 25.4 mm diameter. Resin embedding can be used to prepare irregularly shaped samples (*see* **Note 4**). Our methods will cover the use of silicon wafers. In general, when quantitative isotope ratio data are required, it is advisable to use a 25.4 mm support and keep standards and samples in the center 15 mm of the sample holder.

2.5 NanoSIMS and Analysis

1. CAMECA NanoSIMS 50/L (CAMECA, Gennevilliers, France).
2. OpenMIMS—NRIMS ImageJ Analysis Module. This is an ImageJ plug-in for processing images captured with NanoSIMS 50/L. This plug-in was developed at the National Resource for Imaging Mass Spectrometry (NRIMS) and is available for download on the NRIMS website: http://www.nrims.harvard.edu/software.php.

3 Methods

The basic study design for this experiment is illustrated below (Fig. 2).

3.1 Microbial System

NanoSIMS analysis may be adapted to the study of many different microbial systems. For our study, we are using ^{13}C- and ^{15}N-labelled threonine injected intravenously into host mice to track uptake of

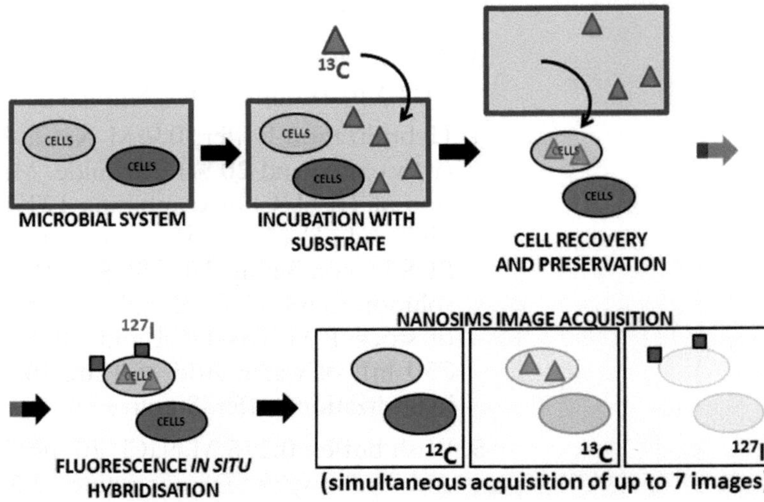

Fig. 2 Basic study design for tracking uptake of substrate in microbial cells by combining in situ hybridization and NanoSIMS analysis. Cells are incubated with ^{13}C-labelled substrate (*triangles*) before recovery and preservation. Halogen-labelled (^{127}I) phylogenetic probes (*squares*) are then hybridized to the cells of interest. Simultaneous imaging of ^{13}C and ^{127}I using NanoSIMS allows mapping of cell metabolic function to identity

the labelled substrate by gut bacteria after 24 h. General considerations for planning an experiment will be outlined in this section:

1. In principle, substrates may be labelled with any rare isotope of an element that exists in the compound to be traced. For maximum sensitivity, universal labelling of an element is desirable; however, any isotopic signal that can be detected above the natural abundance background of control samples is in theory sufficient to answer the question as to whether the organism assimilated the compound. The investigator may wish to only label certain atoms in order to track metabolic processes, but a priori knowledge of the system is likely to be required in this case for meaningful data interpretation. Although radioisotopes may be used in place of stable isotopes, their use is generally not recommended due to the fact that the SIMS instrument may become contaminated to above a regulatory level. For elements with only one stable isotope (e.g., P), the researcher might consider other options for imaging such as microautoradiography:

   (a) Halogens: Halogen atoms are useful for SIMS analysis as they emit a high relative yield of secondary negative ions owing to their high electron affinity and thus can be imaged with exquisite sensitivity. Halogen labels are usually used in the form of deoxyuridines—fluoro-, iodo-, or bromodeoxyuridine. However, care must be taken to account for background levels of halogen atoms. Behrens et al. resolved this issue by combining NanoSIMS analysis with enhanced element labelling (EL)-FISH with the use of halogen-containing fluorescently labelled tyramides as substrates for enzymatic tyramide deposition to boost the halogen signal [14].

2. Incubation:
   (a) Time and temperature of the system: This is dependent on the system used; cultures must be optimized for both the hypothesis tested (test or environmental conditions) and growth of cells and the subsequent uptake of substrate.
   (b) Physicochemical parameters of the substrate: The concentration of label in the substrate chosen must be considered; a high ratio of isotopic label in the compound to be traced will increase signal.
   (c) Substrate delivery: The experimental design should also consider the efficiency of substrate delivery to the target cells. This may be affected by substrate solubility, concentration, system temperature, and pH or natural physical setting in line with the question to be answered. In addition, incubation length should allow sufficient time for uptake of the substrate while minimizing loss from turnover.

3.2 Sample Recovery

After obtaining the target cells, fixation of the cell sample is necessary to halt biochemical activity and preserve cell structure integrity for accurate analysis. Several washing steps are also included to minimize interference from buffer salts and other inhibitors possibly present in the original sample. Microbial cells must also be separated to a sufficient degree as to allow individual characterization while still retaining any cell arrangements. Here, we describe paraformaldehyde fixation of colon contents from culled mice and sample preparation for NanoSIMS analysis. It is important that the experimenter provide non-fixed controls as the paraformaldehyde is certain to dilute the ^{13}C signal present in the cells potentially providing a false negative. Comparison between similar organisms that have been grown in ^{13}C-containing medium with and without fixation should provide an estimate of the dilution factor in ^{13}C signal caused by fixation. Similarly, nucleic acid probes will dilute both ^{13}C and ^{15}N signals. This is potentially more complicated than the case of fixation because the degree of dilution will depend on the number and types of bases contained in the probe and the number of target copies in each cell. Nevertheless, a control can be constructed from a similar organism containing the same target that is harvested in log phase. Such an organism would be expected to contain maximum copies of the target gene if the target resides on a ribosome and thus would serve as a worst-case scenario. An estimate of maximum dilution can thus be made.

3.3 PFA Fixation

1. Homogenize 5 mg fecal sample in ice-cold 1× PBS (1:10; sample:PBS).
2. Centrifuge at 200×*g* for 30 s to remove large fecal particles/debris.
3. Obtain supernatant and centrifuge at 9,000×*g* for 2 min to pellet bacterial cells.
4. Wash thrice with 1× PBS to remove any inhibitory materials.
5. Fix 1 vol sample to 3 vols fresh 4 % (w/v) paraformaldehyde in PBS overnight at 4 °C.
6. Following fixation, centrifuge cells (5 min, 4 °C, 5,000 rev/min).
7. Wash pelleted cells with ice-cold PBS thrice.
8. Resuspend and store in 50 % (v/v) ethanol:PBS at −20 °C until use.

3.4 Additional Labelling: FISH/Phylogeny

1. Wash 10–100 μl fixed cells in 1× PBS to remove ethanol.
2. Resuspend in 1× PBS and mix with an equal volume of 0.02 M EDTA for partial deflocculation.
3. Pellet cells by centrifugation at 11,000×*g* for 2 min.
4. Wash pelleted cells in 400 μl hybridization buffer, and then resuspend in 20 μl preheated hybridization buffer.
5. Add 2 μl of probe (50 ng/μl) and incubate overnight at appropriate temperature.
6. Pellet and wash cells for 15 min in wash buffer at 48 °C.
7. Centrifuge cells for 10 min at 11,000×*g* and resuspend in 50 μl water.
8. Examine cell suspension for density and separation (*see* **Note 5**).

3.5 Sample Preparation, Immobilization, and Coating

If working from cells stored in ethanol:PBS, begin sample preparation from **step 1**. If sample is already in water, proceed directly to **step 2**:

1. Wash 10 μl fixed cells twice in water and resuspend in water (*see* **Note 6**).
2. Spot 1 μl of each washed cell suspensions (including yeast control cells) onto silicon chip (*see* **Note 7**).
3. Allow to air-dry (*see* **Note 8**).
4. Sputter coat cells with 5–10 nm gold (or other conducting metals) to increase sample conductivity.
5. Proceed to NanoSIMS analysis.

3.6 NanoSIMS Analysis

The CAMECA NanoSIMS 50 contains multiple movable detectors that allow the parallel mapping of up to five ion species. Selection of these depends on the experimental design and the signals of interest. In this example, we detect cell uptake of substrate labelled with ^{13}C and ^{15}N:

1. The importance of excellent sample maps cannot be emphasized enough as finding regions of interest using the optical microscope on the NanoSIMS can be nearly impossible otherwise. It is helpful to have fiduciary marks on the sample substrate and to record progressive magnifications of the area of interest with low-magnification images containing the fiduciary mark. In this way, the region of interest can be located relatively easily using a combination of the optical microscope on the NanoSIMS, ion, and/or secondary electron imaging.
2. Position electron multipliers to collect $^{12}C^-$, $^{13}C^-$, $^{12}C^{14}N^-$, and $^{12}C^{15}N^-$ (note that nitrogen cannot be detected directly through N^- and must be analyzed as the CN^- cluster ion and C_2^- ions may give a higher signal from organic matrices than atomic C). The last detector may be positioned such that, for example, $^{16}O^-$, F^-, or $^{28}Si^-$ may be analyzed.
3. Use the CCD (charge-coupled device) camera to find regions on the sample and select an area of interest for image acquisition (*see* **Note 9**), noting the coordinates.
4. Begin by imaging the isotopic standard (yeast cells in this case) to calibrate the detector response before moving on to the test samples (*see* **Note 10**):
   (a) It is important to presputter all standard and analysis areas with the same ion dose prior to analysis. For dispersed cells on a Si substrate, an ion dose of 10^{15} ions/cm^2 works well.
   (b) A defocused beam (EOP lens changed by ~50 V) helps to avoid pitting in the sample surface during presputter and secondary tuning.
   (c) Tune EOS lens and horizontal and vertical centering to obtain a flat response across the imaged area.
   (d) Begin an analysis over the selected area to generate secondary ion images.
5. Repeat image acquisition on samples until enough cells to make a statistically significant conclusion have been imaged (*see* **Note 11**).
6. If halogen-labelled phylogenetic probes have been hybridized to the sample, a second image may be required with a detector moved to the appropriate mass, e.g., m/z 127. The remaining electron multipliers positioned to detect appropriate masses to distinguish cells.
7. Repeat acquisition as before (**steps 1–7**).

3.7 Data Analysis

There are numerous ways to process image data acquired from a NanoSIMS. We prefer to use the OpenMIMS plug-in for ImageJ. ImageJ is a free image analysis software package available for download at http://rsbweb.nih.gov/ij. The OpenMIMS plug-in is available for download at http://www.nrims.harvard.edu/software.php. More comprehensive instructions for using the program are also available online.

3.7.1 Opening the File

1. NanoSIMS image files are saved with two file extension formats—*filename.chk_im* and *filename.im*. The **.chk_im* file stores metadata associated with each image acquisition. The **.im* file stores the binary image data.
2. Open ImageJ > Plugins > OpenMIMS > Open MIMS Image.
3. A new window titled OpenMIMS will open. Go to File > Open MIMS Image and navigate to the appropriate file folder. Select files with the extension .im (*filename.im*). This will open the entire image set covering the masses chosen for visualization of a particular field of view.
4. The "MIMS Data" tab provides information associated with the images captured, including masses detected, dwell time, and raster size.

3.7.2 Generating Hue/Saturation/Intensity (HSI) Images

1. The "Process" tab provides an option to generate HSI images, essentially heatmaps of isotope/elemental ratios as selected by the user (*see* **Note 12**).
2. Click on "Add…" to select ratios of interest to the list. They will appear in the list in the following format: "Mass XX.XX/Mass YY.YY."
3. Highlight the ratio of interest in the list and click on "Display HSI." This will bring up a new window with regions colored according to a heat scale of enrichment.
4. To edit the image, click on the desired window to make it active, and then go to the ImageJ window and select Image from the menu bar to access image editing options.
5. To save, click on the desired window to make it active, and then go to the ImageJ window and select File > Save As and select the desired file format for saving the image.

3.7.3 Selecting Regions of Interest (ROIs)

1. For optimum accuracy and precision, image data must be corrected for dead time, QSA[X], and instrumental mass bias. Although each of these can be achieved on the scale of regions of interest (*see* below) in a spreadsheet, varying pixel brightness for bacterial samples on a support matrix makes correcting images pixel by pixel for dead time and QSA desirable. A macro that can be used in ImageJ to correct for dead time and QSA is provided (*see* **Note 13**).

2. In the OpenMIMS window, select View > ROI Manager. The MIMS ROI Manager window will open.
3. In the ImageJ window, select the Freehand selection tool (fourth along the row of icons below the menu bar).
4. Select a well-resolved image in the set. Using the crosshair cursor, carefully define each separate cell as an ROI by holding down the left mouse button and outlining the cell. Each ROI will be numbered in sequence and can be viewed by highlighting the appropriate number in the ROI list in the MIMS ROI Manager window. ROIs can be moved or hidden by toggling the checkboxes in the lower right side of the window.

3.7.4 Image Analysis

1. After defining all relevant ROIs, click on the first image (the first detected mass) to make it active.
2. Return to the MIMS ROI Manager window and highlight all the ROIs. Click on the "Measure" button on the right side of the window. A new window will appear with the statistics for all ROIs selected for the first image.
3. Highlight the entire dataset and copy it to an Excel sheet.
4. Repeat for the remaining images (detected masses) that are required for analysis.

4 Notes

1. Do not exceed 60 °C—PFA dissociates at 60 °C. A water bath with constantly monitored temperature is useful for this step.
2. Keep the solution heated for this step. Five to ten drops of 10 M NaOH should clear the solution (PFA dissolves at pH 10). Make sure to swirl after each addition to mix solution effectively.
3. Solution should be made up fresh for best results.
4. Resin used for embedding samples must be carefully prepared and cured as some resins will degas over time. The amount of resin used can be reduced by minimizing disc thickness or including metal rings. Recommended resin formula: Combine 502.26 g Araldite with 24 g dodecenyl succinic anhydride (DDSA) thoroughly, avoiding bubbles. Stand the mixture for 30 min to allow and bubbles to disappear. Add 1.4 g benzyldimethylamine (BDMA) and mix thoroughly, avoiding bubbles. Embed samples as required and cure at 60 °C for at least 24 h. Samples can then be microtomed and sections mounted on silicon wafers or metal discs.
5. Cell density should be examined by basic light microscopy to ensure a clump-free, even dispersion of cells throughout the suspension. If using fluorescent-conjugated probes, hybridized

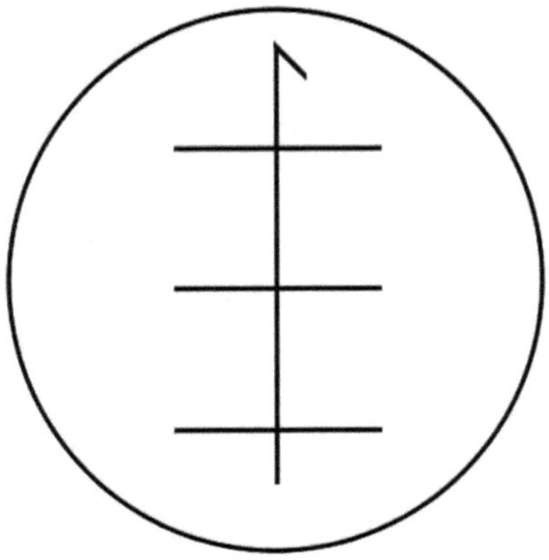

Fig. 3 Etching the solid support with an arrow and grid allows sample orientation and mapping for analysis

samples can be examined under fluorescence microscopy (and compared with positive and negative control cells) to determine labelling efficiency.

6. Buffer salts must be removed from the sample to avoid contaminating analysis. Dilute further if required to reach suitable cell density for imaging and analysis.
7. It is helpful to mark out sectors/grid on the solid support to aid orientation during NanoSIMS analysis and to allow for pre-mapping of the sample with techniques such as fluorescence microscopy (Fig. 3).
8. The NanoSIMS analysis is done in a high-pressure vacuum (10^{-10} Torr), and care must be taken to remove all water from the sample. We found air-drying to be sufficient for our purposes, but experimental aim must be considered when conducting sample preparation to maintain sample preservation. Methods that are commonly used are fast freezing, low-temperature dehydration, chemical fixation, and resin embedding.
9. The CCD camera allows indirect optical imaging of the sample. It is useful to orient sample placement based on the sectors marked out on the sample support (silicon chip). It also allows a quick scan for cell separation/density and identification of areas of interest for NanoSIMS image acquisition.
10. The native ratio of ^{13}C to ^{12}C is ~1 %, and the ^{15}N:^{14}N ratio is ~0.37 %. The yeast standard cells and the unlabelled control experimental cells should return a similar value within uncertainty.
11. The number of cells required depends on the hypothesis being tested. 80–120 cells were imaged per sample for the study from which this method is taken. The number of images required will vary depending on cell density and raster size.

12. HSI images are produced by combining the ratio value of a pixel, counts of one of the selected masses for intensity, and a constant saturation value to generate pixels in RGB color space.
13. The following is an example of an ImageJ macro that corrects for dead time and QSA assuming a primary stage current of 2.7 pA, a 44 ns dead time, and a QSA coefficient of 0.5[X] and a 32 bit counting card on the NanoSIMS:

```
//DEADTIME AND QSA ONE IMAGE FRAME
run("32-bit");
//image dwell time in microseconds
dwell=27500
//Primary beam current in pA
Ip=2.7//iterate over all pixels
for (y=0; y<getHeight(); y++)
for (x=0; x<getWidth(); x++) {
value=getPixel(x, y);
//evaluate the formula
                  v1=value*1/dwell*1e6;
v2=v1/(1-44e-9*v1);
                  v3=v2/1e6*dwell;
   v4=v3/(Ip*6241509);
   v5=v4/(1-v4/2);
   v6=v3*(1+v5/2);
setPixel(x, y, v6);
}
```

Acknowledgement

The authors received support from a Travel and Access Program (TAP) grant from the Australian Microscopy and Microanalysis Research Facility (AMMRF).

References

1. Castaing R, Slodzian Z (1962) Microanalyse per emission ionique secondaire. J Microsc 1:395–410
2. Benninghoven A (1973) Surface investigation of solids by the statical method of secondary ion mass spectroscopy (SIMS). Surf Sci 35:427–437
3. Levi-Setti R, Crow G, Wang YL (1985) Progress in high resolution scanning ion microscopy and secondary ion mass spectrometry imaging microanalysis. Scan Electron Microsc (pt2): 535–552
4. Migeon HN, Schuhmacher M, Slodzian G (1990) Analysis of insulating specimens with the Cameca IMS 4f. Surf Interface Anal 16:9–13
5. Slodzian G et al (1992) Scanning secondary ion analytical microscopy with parallel detection. Biol Cell 74:43–50

6. Messenger S et al (2004) Samples of stars beyond the solar system: silicate grains in interplanetary dust. Science 300:105–108
7. Floss C et al (2004) Carbon and nitrogen isotopic anomalies in an anhydrous interplanetary dust particle. Science 303:1355–1358
8. Lozano-Perez S et al (2009) Multi-scale characterization of stress corrosion cracking of cold-worked stainless steels and the influence of Cr content. Acta Mater 57:5361–5381
9. Clode PL et al (2009) In situ mapping of nutrient uptake in the rhizosphere using nanoscale secondary ion mass spectrometry. Plant Physiol 151:1751–1757
10. Guerquin-Kern J-L et al (2004) Ultrastructural cell distribution of the melanoma marker iodobenzamide: improved potentiality of SIMS imaging in life sciences. Biomed Eng Online 3:1–7
11. Lechene C et al (2007) Quantitative imaging of nitrogen fixation by individual bacteria within animal cells. Science 317:1563–1566
12. Popa R et al (2007) Carbon and nitrogen fixation and metabolite exchange in and between individual cells of *Anabaena oscillarioides*. ISME J 1:354–360
13. Finzi-Hart JA et al (2009) Fixation and fate of C and N in the cyanobacterium *Trichodesmium* using nanometer-scale secondary ion mass spectrometry. Proc Natl Acad Sci U S A 106: 6345–6350
14. Behrens S et al (2008) Linking microbial phylogeny to metabolic activity at the single-cell level by using enhanced element labeling-catalyzed reporter deposition fluorescence in situ hybridization and nanometer-scale secondary-ion mass spectrometry. Appl Environ Microbiol 74:3143–3150
15. Li T et al (2007) Simultaneous analysis of microbial identity and function using NanoSIMS. Environ Microbiol 10:580–588
16. Woebken D et al (2012) Identification of a novel cyanobacterial group as active diazotrophs in a coastal microbial mat using NanoSIMS analysis. ISME J 6:1427–1439
17. Musat N et al (2008) A single-cell view on the ecophysiology of anaerobic phototrophic bacteria. Proc Natl Acad Sci U S A 105:17861–17866
18. Fike DA et al (2008) Micron-scale mapping of sulfur cycling across the oxycline of a cyanobacterial mat: a paired nanoSIMS and CARD-FISH approach. ISME J 2:749–759

Chapter 12

Single-Cell Raman Sorting

Mengqiu Li, Dan G. Boardman, Andrew Ward, and Wei E. Huang

Abstract

Single-cell Raman spectroscopy is a noninvasive and label-free technology for biochemical analysis of bacterial cells. A single-cell Raman spectrum functions as a metabolic "fingerprint" of an individual cell, which enables differentiation of cell types, physiological states, nutrient condition, and variable phenotypes. Raman tweezers combines single-cell Raman spectroscopy with optical laser tweezers to allow the identification and isolation of single living cells according to their Raman spectra. After cell sorting subsequent culturing and genomic sequencing has the potential to reveal totally new groups of microbes. Stable isotope probing with Raman tweezers offers a culture-independent toolbox to study genetic functions and physiology of unculturable microorganisms in the ecosystem.

Key words Raman, Unculturable microorganisms, Single cell, Raman sorting, Stable isotope probing

1 Introduction

It is estimated that over 99 % of microorganisms are, as yet, unculturable using standard microbiology techniques [1–7]. To date, metagenomics with high-throughput sequencing is the major culture-independent approach for studying unculturable microbes. The approach circumvents the cultivation issue by extracting the total DNA from a microbial community environment to enable sequencing analysis and functional studies. However, analysis of genetic information cannot replace the cell-based analysis for the study of microbial physiology and phenotype. To study the microbial community and to recover the functional genes harbored by previously uncultured microorganisms, it is crucial that one can characterize and recover whole single cells from complex samples.

Fluorescence-activated cell sorting (FACS) is a powerful tool to separate cell populations. However, FACS characterizes and sorts cells by excitation of a fluorescent label which has to be chemically linked to the cells as most bacteria have weak or no fluorescence. In addition, fluorescent-staining-based FACS is confined to measure 6–10 simultaneous parameters and provides limited information on

Ian T. Paulsen and Andrew J. Holmes (eds.), *Environmental Microbiology: Methods and Protocols*, Methods in Molecular Biology, vol. 1096, DOI 10.1007/978-1-62703-712-9_12, © Springer Science+Business Media, LLC 2014

cellular phenotype and metabolic states. Raman microspectroscopy is a noninvasive label-free technology which can provide an intrinsic chemical "fingerprint" of a single bacterial cell. Raman microspectroscopy detects vibrations of chemical bonds of molecules through the inelastic scattering of incident laser light [8]. A typical Raman spectrum embodies a rich cellular chemical profile, including >1,000 Raman bands covering nucleic acids, protein, carbohydrates, and lipids, which enables the characterization of different cell types and physiological and phenotypic changes to living cells [8]. Raman tweezers integrates Raman single-cell spectroscopy and optical laser tweezers to simultaneously measure and sort bacteria at single-cell level. In combination with stable isotope probing (SIP), Raman tweezers enables the identification and isolation of key microbes which carry out specific ecological functions, e.g., CO_2 fixation and contaminant biodegradation. The direct Raman spectroscopic measurement of cellular metabolism in living uncultured microbes has the potential to provide much needed information about the ecological role of microbial species. Vibrational spectroscopic techniques, such as Raman spectroscopy, are sensitive to stable isotope in molecules, for example, single cells which incorporate C^{13} or N^{15} into their cellular structures are easily detected by frequency shifts in the Raman spectral signals [9–12].

In the application of Raman tweezers to cell sorting, single cells can be captured in a liquid sample using an optical tweezers configuration (e.g., 488, 514.5, 532, 633, 785, or 1,064 nm laser), and then a Raman measurement laser (which is commonly the same laser) is used as the spectroscopic light source for Raman spectral acquisition. Single-cell Raman spectrum (SCRS) from cells containing ^{12}C has a sharp band at ~1,002 cm^{-1}, while SCRS from cells containing C^{13} would show a red-shifted band at ~966 cm^{-1}. This sharp Raman band can be used as a biomarker for differentiating cells that incorporate ^{13}C from a ^{13}C substrate (e.g., CO_2), and the cells can be sorted and isolated by optical tweezers manipulation to permit attempted culturing and genomic sequencing. It has been shown that with careful control of the laser power and wavelength, sorted single microbial cells maintain their viability after the procedure and cells are suitable for single-cell genome amplification [12].

2 Materials

2.1 Raman Sorting System

The Raman sorting system used in these studies was constructed on a Leica DM-IRB microscope equipped with a water immersion objective (63×, NA = 1.2, Leica, UK, HCX PL APO) by combining a Raman spectrometer and a dual-wavelength 514.5 nm and 1,064 nm laser tweezers apparatus (Laser 2000, UK, LCS-DTL-322; Coherent, CA, USA, Innova 90-5 Ar-ion) (*see* **Note 9**).

Raman spectra were acquired in a backscattered arrangement using a 514.5 nm laser where the laser line was removed by an edge filter (Semrock, IL, USA, RazorEdge). A 0.5 m spectrograph with 1,200 g/mm grating (Acton, MA, USA, SP 2500i) and a liquid nitrogen-cooled CCD (Princeton Instruments, USA, Spec 10:400 BR DD) were used to provide single-cell Raman spectra with a resolution of 0.85 rel/cm per CCD pixel. The 514.5 nm and 1,064 nm lasers were directed in separate optical pathways and combined by a dichroic mirror to enable superimposition of the focal positions by the microscope objective (*see* **Notes 1** and **2**). A stepper motor microscope translation stage was joystick controlled and also programmed using LabVIEW software to move to user-defined positions at controlled velocities (*see* **Notes 8–10**).

2.2 Consumable Items and Other Materials

1. Capillary tubes (Camlab, UK, VD/5010-050) (*see* **Note 4**).
2. Single-cell genome amplification kits (Qiagen, UK, REPLI-g Mini Kit).
3. PCR primers: ITS4 and ITS5 for yeast [13] and 63F and 1387R for bacteria [14].
4. A thermal cycler for PCR reactions.
5. pGEM-T vectors (Promega, UK) and *Escherichia coli* strain JM109.
6. An agarose gel tank and a voltage controller.
7. A UV or blue light gel imaging dock.
8. Access to DNA sequencing facility.
9. A fluorescence microscope.
10. Sterilized microcentrifugation tubes and Erlenmeyer flasks.
11. Chemicals and growth media: 70 % ethanol, sterilized 0.85 % NaCl saline, sterilized L broth (LB), sterilized LB with 50 μg/ml kanamycin, and sterilized minimal medium [15] supplemented with 30 mM ^{12}C glucose or ^{13}C glucose as needed.

3 Methods

3.1 Microorganism Preparation

The protocol below describes an example control experiment to separate bacterial cells utilizing glucose from two other microbial cells. The basic protocol can be used to isolate a single cell from a mixture based on its utilization of a ^{13}C-labelled carbon source. Yeast strain *Saccharomyces cerevisiae* was incubated in LB medium at 28 °C overnight. *Escherichia coli* strain DH5α was incubated in minimal medium supplemented with ^{12}C glucose at 37 °C overnight. *Pseudomonas fluorescens* strain SBW25::Km-RFP was incubated in minimal medium supplemented with ^{13}C glucose at 28 °C overnight. *P. fluorescens* strain SBW25::Km-RFP has a kanamycin-resistance gene and red fluorescent protein (RFP) gene inserted into the

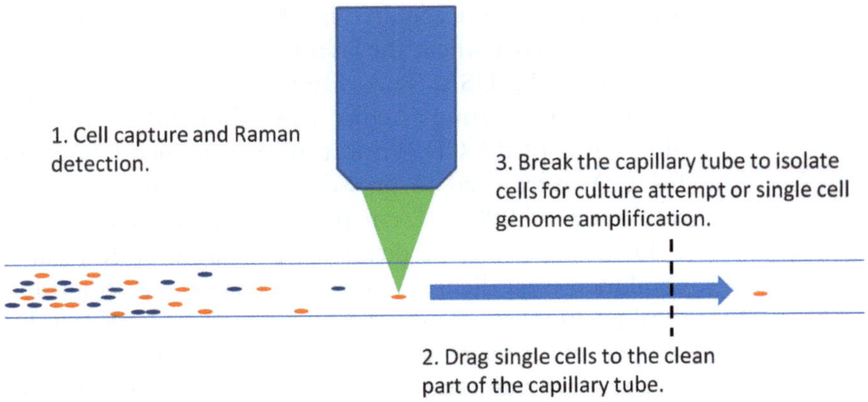

Fig. 1 Schematic diagram of the Raman cell sorting system

chromosome using modified mini-Tn5 [16], which could be used to confirm identities of sorted cells. All cells were collected by centrifugation and washed by NaCl solution three times (*see* **Note 7**). Cells were also diluted down to approximately 10^5 cells/ml level (*see* **Note 5**). *E. coli* strain DH5α cells and *P. fluorescens* strain SBW25::Km-RFP cells were mixed evenly for sorting.

3.2 Raman Sorting

Yeast cell suspension and mixed bacterial cell suspensions were transferred into autoclaved capillary tubes by the following method: dipping one end of an autoclaved capillary tube into sterile distilled water to fill about two thirds of its length, then dipping the same end again into the cell suspension to fill the capillary tube (*see* **Note 3**). Filled capillary tubes were then wiped with 70 % ethanol and fixed onto a microscope slide by using Vaseline to seal both ends.

Capillary tube-slide assembly is then mounted on the microscope translation stage. A single yeast or bacterial cell is then captured by the focused 1,064 nm infrared laser beam (5 mW) and dragged to an area containing fewer cells to gather Raman signal without interference from surrounding cells. The 514.5 nm laser (9 mW) is then unshuttered for Raman spectrum acquisition with integration time of 30 s for each single cell. Selected single cells are then moved again by the 1,064 nm laser beam to the clear part of the capillary tube (approx 1 cm away). The capillary tube is then removed from the slide and broken, and the part that contained the isolated cell was transferred to a microcentrifugation tube for single-cell genome amplification or a flask for cell growth as shown in Fig. 1.

For carbon utilization experiments such as this, the decision to release a bacterial cell or to move it for isolation is based on the "red shift" of the substituted benzene ring derivatives. This Raman band is usually at 1,002 cm^{-1} but shifted to 966 cm^{-1} when the organism is grown in ^{13}C media, as shown in Fig. 2. *P. fluorescens* strain SBW25::Km-RFP cells are identified by this distinctive band shift and isolated as described above (*see* **Notes 6** and **10**).

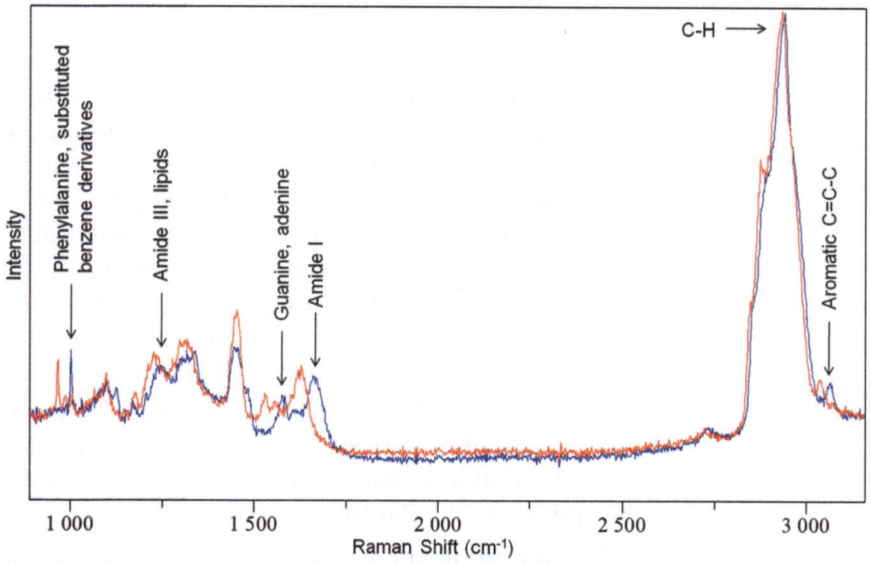

Fig. 2 Raman spectra of single bacterial cell showing the shift of bands caused by stable isotope labelling

Due to possibility of cell diffusion and contamination from external source, it is advisable to have control capillary tubes. Control capillary tubes containing the cell suspensions are prepared and broken in the same way but without cell manipulation into the cell-free region. In order to assess the effect of cell diffusion in the capillary tubes, control tubes are left by the same length of time before being broken.

3.3 Checking Results and Further Analysis

Together with the control capillary tubes the isolated yeast cells and the isolated SBW25 cells were transferred into flasks to incubate in LB at 28 °C and LB-kanamycin broth at 28 °C, respectively. The size and morphology of yeast cells make them useful markers of successful Raman sorting because they are readily distinguished from most microorganisms present in the laboratory environment which could contaminate the capillary tube. The viability of *P. fluorescens* strain SBW25::Km-RFP in LB-kanamycin medium and their red fluorescence were used as the markers of successful Raman sorting in this instance.

Some isolated yeast cells, SBW25 cells, and control samples were transferred to microcentrifugation tubes and subject to single-cell genome amplifications using REPLI-g Mini Kits. Confirmation of the identity of recovered cells is performed by ribosomal RNA sequence analysis. PCR is used to recover the ribosomal RNA gene sequences from the resultant amplified genome DNA, using primer pairs ITS4/ITS5 for yeast [13] or 63F/1387R for bacteria [14]. The PCR products are cloned into pGEM-T vectors and transferred to *Escherichia coli* strain JM109 cells for DNA sequencing.

4 Notes

1. The Raman tweezers apparatus requires careful alignment to enable laser tweezers capture, clear focused microscope images of the bacteria, optimum acquisition rates of Raman spectra, and easy transfer from 1,064 nm trapping to 514.5 nm trapping. Alignment optimization is performed using a dilute 1 μm diameter polystyrene bead dispersion (Interfacial Dynamic Corporation) that is prepared in a separate capillary tube from which a single bead is trapped and used to optimize the parameters described above.
2. The illumination light source of the microscope is filtered to only allow light of wavelengths below 500 nm onto the sample. This enables simultaneous visualization of the sample and acquisition of the Raman spectral signals.
3. When preparing the cell samples in the capillary tubes, it is important to minimize mixing of the water and cell dispersion. Inspection of the interface between these regions should show a transition length of no more than 300 μm from the clean to cell-containing areas.
4. Capillary tubes should be soaked in detergent and then rinsed by water. After that, the capillary tubes are dried at 105 °C oven overnight.
5. To isolate bacteria that have flagella, Ficoll (7.5 % v/v) can be added in suspension medium to increase viscosity and thus inhibit cellular mobility.
6. While several peaks in the Raman signal shift with increasing ^{13}C content, in practice we find the phenylalanine peak at 1,002 cm^{-1} as the most useful primary marker since it could often be distinguished with acquisition times of only a second.
7. When sorting multiple bacteria, it was observed that the collected non-flagellated cells remained within 50 μm of the drop-off point. If a quick check was required of the isolated species, this could be done easily.
8. The translation velocity of the stage was optimized to be as fast as possible without losing the bacteria (through viscous drag forces). While the stage could be stopped and reversed to rescue the dropped bacteria, this is a time-consuming process. The velocity of the stage is typically between 100 and 200 μm/s. In the described experiment, all translations were performed using the 1,064 nm laser only.
9. The Raman sorting system described in this method is an example of possible configurations. Other Raman microscopes and different lasers can be used provided that they can acquire single-cell Raman spectra within short time (30 s or shorter) and manipulate single cells without damaging their viability.

10. Any microorganisms with distinguishable Raman signature can be used for single-cell Raman sorting. Other types of analysis suitable for single cells can also be combined with Raman sorting technique.

Acknowledgements

We thank Dr. Simon FitzGerald in HORIBA Scientific Ltd. for his technical input and helpful discussion. We are indebted to the Central Laser Facility at the Rutherford Appleton Laboratory for providing access to the instrument.

References

1. Wagner M et al (1993) Probing activated-sludge with oligonucleotides specific for proteobacteria—inadequacy of culture-dependent methods for describing microbial community structure. Appl Environ Microbiol 59: 1520–1525
2. Riesenfeld CS et al (2004) Metagenomics: genomic analysis of microbial communities. Annu Rev Genet 38:525–552
3. Jones JG (1977) Effect of environmental-factors on estimated viable and total populations of planktonic bacteria in lakes and experimental enclosures. Freshwater Biol 7: 67–91
4. Ferguson RL et al (1984) Response of marine bacterioplankton to differential filtration and confinement. Appl Environ Microbiol 47:49–55
5. Torsvik V et al (1990) High diversity in DNA of soil bacteria. Appl Environ Microbiol 56: 782–787
6. Ellis RJ et al (2003) Cultivation-dependent and -independent approaches for determining bacterial diversity in heavy-metal-contaminated soil. Appl Environ Microbiol 69:3223–3230
7. Rappe MS, Giovannoni SJ (2003) The uncultured microbial majority. Annu Rev Microbiol 57:369–394
8. Huang WE et al (2010) Shining light on the microbial world: the application of Raman microspectroscopy. Adv Appl Microbiol 70: 153–186
9. Huang WE et al (2009) Resolving genetic functions within microbial populations: in situ analyses using rRNA and mRNA stable isotope probing coupled with single-cell Raman-fluorescence in situ hybridization. Appl Environ Microbiol 75:234–241
10. Huang WE et al (2004) Raman microscopic analysis of single microbial cells. Anal Chem 76:4452–4458
11. Huang WE et al (2007) Raman-FISH: combining stable-isotope Raman spectroscopy and fluorescence in situ hybridization for the single cell analysis of identity and function. Environ Microbiol 9:1878–1889
12. Huang WE et al (2009) Raman tweezers sorting of single microbial cells. Environ Microbiol Rep 1:44–49
13. Liberal ATD et al (2005) Contaminant yeast detection in industrial ethanol fermentation must by rDNA-PCR. Lett Appl Microbiol 40: 19–23
14. Marchesi JR et al (1998) Design and evaluation of useful bacterium-specific PCR primers that amplify genes coding for bacterial 16S rRNA. Appl Environ Microbiol 64: 795–799
15. Huang WE et al (2005) Chromosomally located gene fusions constructed in Acinetobacter sp ADP1 for the detection of salicylate. Environ Microbiol 7:1339–1348
16. Delorenzo V et al (1990) Mini-Tn5 transposon derivatives for insertion mutagenesis, promoter probing, and chromosomal insertion of cloned DNA in gram-negative eubacteria. J Bacteriol 172:6568–6572

Chapter 13

Bacterial Whole-Cell Biosensors for the Detection of Contaminants in Water and Soils

Yun Wang, Dayi Zhang, Paul A. Davison, and Wei E. Huang

Abstract

Bacterial whole-cell biosensors (BWBs) have unique advantages over conventional environmental monitoring techniques on the detection of toxicity and bioavailability of contaminants in water and soils. BWBs can also be rapid, sensitive, semiquantitative, cost-effective, and easy to use. In this study, a standard method is described for the detection of contaminants and toxicity in real water and soil samples using *Acinetobacter baylyi* ADP1-based biosensors.

Key words Biosensor, *Acinetobacter baylyi* ADP1, Bioluminescence, Genotoxicity, Environmental samples

1 Introduction

Modern developments in industry and agriculture have resulted in an increased use of fossil fuels, pesticides, and metals. As a consequence, contaminants such as polycyclic aromatic hydrocarbons (PAHs), organophosphorus pesticides, and heavy metals (e.g., Hg, Pb) have been released into the environment, seriously threatening both natural ecosystems and human health. The detection of toxicity and bioavailability of contaminants in the environment using traditional chemical analysis is laborious and costly [1, 2]. In particular, emergency incidents such as the recent 2010 oil spill in the Gulf of Mexico demand a rapid environmental risk assessment which includes chemical detection, toxicity, and bioavailability [3]. Bacterial whole-cell biosensors (BWBs) could provide a rapid and semiquantitative estimation of specific compounds and an assessment of their toxicity and ecological impact on the environment [4, 5].

Acinetobacter baylyi ADP1 is a nutritionally versatile chemoheterotrophic bacterium found in a wide range of water and soil environments. A series of ADP1-based BWBs have been constructed, and several of these have been found to be robust and reliable with an accuracy comparable to GC-MS detection [6]. Besides their unique advantages in detection of bioavailability and

Ian T. Paulsen and Andrew J. Holmes (eds.), *Environmental Microbiology: Methods and Protocols*, Methods in Molecular Biology, vol. 1096, DOI 10.1007/978-1-62703-712-9_13, © Springer Science+Business Media, LLC 2014

toxicity, these BWBs have several other advantages [7, 8]: (1) the detection operation is straightforward and easy for water and soil samples with minimal pretreatment; (2) the semiquantitative detection of contaminants takes less than 2 h, considerably quicker than US EPA methods, some of which take several days; (3) the detection can be highly specific and sensitive to specific compounds; (4) the detection range can be of 3–6 orders of magnitude (e.g., nM to mM); and (5) it enables high-throughput and low-cost detection.

Here, three *Acinetobacter baylyi* ADP1-based biosensors were employed for sensing specific chemical compounds in environmental samples, namely, salicylate (ADPWH_lux) [9], toluene/xylene (ADPWH_Tol) [10] and n-alkane/crude oil (ADPWH_alk) [11, 12], and one was used for detecting general toxicity (ADPWH_recA) [13]. In this study, we describe the standard methodology for BWBs detection of contaminants in real environmental water and soil samples.

2 Materials

2.1 Reagents

All the deionized water used for solution preparation has a sensitivity of 18 MΩ cm at 25 °C.

Unless otherwise stated, all reagents are analytical grade reagents and are purchased from Sigma-Aldrich, UK. Prepare and store all reagents at room temperature (unless otherwise specified).

1. Bauchop and Elsden solution: dissolve 10.75 g $MgSO_4$, 4.5 g $FeSO_4 \cdot 7H_2O$, 2.0 g $CaCO_3$, 1.44 g $ZnSO_4 \cdot 7H_2O$, 1.12 g $MnSO_4 \cdot 4H_2O$, 0.25 g $CuSO_4 \cdot 5H_2O$, 0.28 g $CoSO_4 \cdot 7H_2O$, 0.06 g H_3BO_3, and 51.3 mL concentrated HCl solution in deionized water, adjust to 1 L [14] (*see* **Note 1**).
2. Luria-Bertani (LB) medium: dissolve 25.0 g LB Broth (Merck, Germany) in deionized water, adjust to 1 L (*see* **Note 1**).
3. LB agar: dissolve 40 g LB agar powder in deionized water, adjust to 1 L (*see* **Note 1**).
4. Minimal medium (MM): dissolve 2.5 g Na_2HPO_4, 2.5 g KH_2PO_4, 1.0 g NH_4Cl, 0.1 g $MgSO_4 \cdot 7H_2O$, 10 μL saturated $CaCl_2$ solution, 10 μL saturated $FeSO_4$ solution, and 1 mL Bauchop and Elsden solution in deionized water, adjust to 1 L (*see* **Note 1**).
5. Sodium succinate solution (1 M): dissolve 162.0 g sodium succinate in deionized water, adjust to 1 L (*see* **Note 2**).
6. Minimal medium (MM) with succinate (MMS): mix 20 mL sodium succinate solution (1 M) with 980 mL MM. Final sodium succinate concentration is 20 mM.

7. Sodium chloride solution (0.85 %): dissolve 8.5 g NaCl in deionized water, adjust to 1 L (*see* **Note 1**).
8. Chloroform ($CHCl_3$).
9. Sodium salicylate ($NaC_7H_5O_3$).
10. Toluene (C_7H_8).
11. Crude oil (from Brent reservoir, North Sea, UK).
12. Mitomycin C (MMC, $C_{15}H_{18}N_4O_5$).

2.2 Equipment

1. 96-well black, clear bottom microplate (e.g., Costar, Corning Inc., USA).
2. 96-well white, clear bottom microplate (e.g., Costar, Corning Inc., USA).
3. −80 °C freezer (e.g., DF8517; ilshin Lab Co. Ltd, USA).
4. −20 °C freezer (e.g., RFA52; Hotpoint, UK).
5. Shaking incubator (e.g., Model G25; New Brunswick Scientific, USA).
6. Centrifuge (e.g., 5415R and 5804; Eppendorf, Germany).
7. Multimode microplate reader (e.g., Synergy 2; BioTek Instruments, Inc., USA).
8. Ultrasonic Bath (e.g., LNGF175; Langford Electronics Ltd, UK).
9. Centrifuge tubes (1.5 mL, sterile, e.g., Eppendorf, Germany).
10. Centrifuge tubes (50 mL, sterile, e.g., Eppendorf, Germany).
11. 1 L bottle (sterile, e.g., Fisherbrand, UK).
12. 50 mL bottle (sterile, e.g., Fisherbrand, UK).
13. 5 mL bottle (sterile, e.g., Fisherbrand, UK).
14. Petri dish (sterile).
15. 0.20 μm disposable sterile filter (e.g., Millipore, USA).
16. Benchtop vortexer (e.g., Labnet International Inc., USA).
17. Bunsen burner.

2.3 Biosensor Strains

Bacterial strains required for this method are available by request from the authors. Send requests to Dr. Wei Huang, Kroto Research Institute, Department of Civil and Structural Engineering, Sheffield University, Broad Lane, Sheffield, S3 7HQ, UK.

1. *Acinetobacter sp.* ADPWH_lux [9] uses *Acinetobacter baylyi* ADP1 as the host and *luxCDABE* as the reporter gene, with expression dependent on the *salA–salR* regulatory system.
2. *Acinetobacter sp.* ADPWH_alk [11] uses *Acinetobacter baylyi* ADP1 as the host and *luxCDABE* as the reporter gene, with expression dependent on the *alkM–alkR* regulatory system.

3. *Acinetobacter sp.* ADPWH_recA [13] uses *Acinetobacter baylyi* ADP1 as the host and *luxCDABE* as the reporter gene, with expression dependent on the *recA* regulatory system.
4. *Acinetobacter sp.* ADPWH_Tol [10] uses *Acinetobacter baylyi* ADP1 as the host and *luxCDABE* as the reporter gene, with expression dependent on the *pu–xylR* regulatory system.

3 Methods

Carry out all procedures at room temperature unless otherwise specified.

3.1 Preparation and Storage of Biosensors

1. Streak out cells of the *Acinetobacter* biosensors stored at −80 °C (*see* **Note 3**) onto LB agar plates and incubate at 30 °C overnight (*see* **Note 4**).
2. Pick a single colony from the agar plate, inoculate 10 mL liquid LB medium, and incubate at 30 °C overnight (*see* **Note 4**).
3. Wash the cells three times with 0.85 % NaCl solution by centrifugation at 900 × *g* for 10 min (*see* **Note 5**) and resuspend the pellet in 10 mL sterile deionized water to create the biosensor stock solution (*see* **Note 6**).
4. Store the biosensor stock solution at 4 °C (*see* **Note 7**).
5. Take 1 mL biosensor stock solution (except for ADPWH_recA), centrifuge at 900 × *g* for 10 min (*see* **Note 5**), and resuspend in 1 mL mineral salts medium with succinate (MMS) to make the biosensor reaction solution prior to detection.
6. Take 1 mL ADPWH_recA stock solution, centrifuge at 900 × *g* for 10 min (*see* **Note 5**), and resuspend in 10 mL LB medium to make the ADPWH_recA reaction solution prior to detection.

3.2 Sample Pretreatment

3.2.1 Water Sample Pretreatment

1. Homogenize a 2 mL sample using 40 kHz ultrasound for 300 s in a 5 mL bottle (*see* **Note 8**).
2. Transfer 1 mL homogenized sample into a 1.5 mL microcentrifuge tube and centrifuge at 4,000 × *g* for 10 s (*see* **Note 9**).
3. Remove 800 μL of the supernatant and keep for detection.

3.2.2 Water Standards Preparation

1. Add an appropriate mass of target pollutants to give a stock concentration of 100 mg/L into 20 mL of a fresh non-contaminated negative water sample in a 50 mL bottle (*see* **Note 10**).
2. Homogenize the mixture using 40 kHz ultrasound in a 50 mL bottle for 300 s and mix well by vortexing for 10 s.
3. Transfer 1 mL of homogenized sample into a 1.5 mL microcentrifuge tube and centrifuge at 4,000 × *g* for 10 s (*see* **Note 9**).

4. Remove 800 μL of the supernatant and keep as the standard pollutant stock solution with a concentration of 100 mg/L.
5. Dilute the standard pollutant stock solution using non-contaminated water sample to make a standard series containing a range of concentrations (*see* **Note 11**).

3.2.3 Soil Sample Pretreatment

1. Take four subsamples from each soil sample weighing 10.0, 50.0, 100.0, and 200.0 mg (*see* **Note 12**).
2. Mix the subsamples with 5 mL sterilized deionized water in a 20 mL bottle.
3. Homogenize the mixture using 40 kHz ultrasound for 300 s (*see* **Note 8**).
4. Transfer 1 mL of homogenized subsample into a 1.5 mL microcentrifuge tube and centrifuge at 4,000×*g* for 10 s (*see* **Note 9**).
5. Remove 800 μL of the supernatant and keep for detection.

3.2.4 Soil Standards Preparation

1. Add appropriate weights of the standard crude oil to be tested to 10 mL chloroform to obtain a concentration range of 0.0 %, 0.12 %, 0.24 %, 0.60 %, 1.2 %, 3.0 %, 6.0 %, 12.0 %, and 24.0 % (v/v) and mix well by vortexing.
2. Transfer each mixture into 10.0 g of standard fresh negative soil and homogenize using 40 kHz ultrasound for 30 s and mix well by vortexing.
3. Volatilize the oil–chloroform–soil mixtures at 30 °C for 48 h until no chloroform residue is present in the soil so that the standard soil samples will contain crude oil with contents of 0.0 %, 0.1 %, 0.2 %, 0.5 %, 1.0 %, 2.5 %, 5.0 %, 10.0 %, 15.0 %, and 20.0 % (v/v).
4. Take four subsamples from each soil sample weighing 10.0, 50.0, 100.0, and 200.0 mg (*see* **Note 12**).
5. Mix the subsamples with 5 mL sterilized deionized water in a 20 mL bottle.
6. Homogenize the mixture using 40 kHz ultrasound for 300 s (*see* **Note 8**).
7. Transfer 1 mL of homogenized subsample into a 1.5 mL microcentrifuge tube and centrifuge at 4,000×*g* for 10 s (*see* **Note 9**).
8. Remove 800 μL of the supernatant and keep for detection.

3.3 Bioluminescence Detection

1. Transfer 1.0 mL of biosensor reaction solution into a 1.5 mL microcentrifuge tube and incubate at 30 °C for 10 min (*see* **Note 13**).
2. Add 20 μL biosensor reaction solution (except for ADPWH_recA) into each well of a 96-well black, clear bottom microplate (*see* **Note 14**).

3. Add 180 μL of the test samples into the biosensor reaction solution containing wells and include a well containing 180 μL of deionized water as a negative control.
4. Apply at least three replicates for each sample (four replicates are recommended).
5. For genotoxicity detection, add 180 μL of ADPWH_recA reaction solution and 20 μL test sample into each well.
6. Place the microplate into the microplate reader and incubate at 30 °C for 4 h.
7. Measure the bioluminescence and optical density at 600 nm (OD_{600}) every 5 min, with 1 min shaking before each reading (*see* **Note 15**).

3.4 Data Analysis

1. An average of three or four replicates taken at the same time forms the bioluminescence and OD_{600} curves for each sample. Calculate cell number at each time point using a standard equation relating OD_{600} to cell count (*see* **Note 16**).
2. The bioluminescence value is divided by the cell count to give the relative bioluminescence; the ADPWH_lux response to a concentration series of salicylate acid is shown in Fig. 1a as an example.
3. Calculate bioluminescence response ratios by dividing bioluminescence of induced samples by that of corresponding controls (non-induced) at the same time point, as shown in Fig. 1b.
4. The relative bioluminescence response ratios of ADPWH_lux, ADPWH_Tol, and ADPWH_alk are obtained by averaging five sets of bioluminescence data monitored around the peak time (*see* **Note 17**), while that of ADPWH_recA is gained by averaging five sets of bioluminescence data monitored between 180 min and 210 min.
5. Establish calibration curves of the relative bioluminescence response ratio against the dose of specific biosensors. The recommended calibration curves of pollutants in pure water are shown in Fig. 2.
6. Compare the relative bioluminescence response ratio of the samples to the calibration curve to obtain the concentration of the pollutants.

As an example to show how contaminants in environmental samples are calculated in practice, an artificial and a real environmental water sample were measured and evaluated using whole-cell biosensors, following the procedure mentioned above. The artificial water sample contains specific amounts of inducers in pure water, where the final concentrations of salicylate acid, toluene,

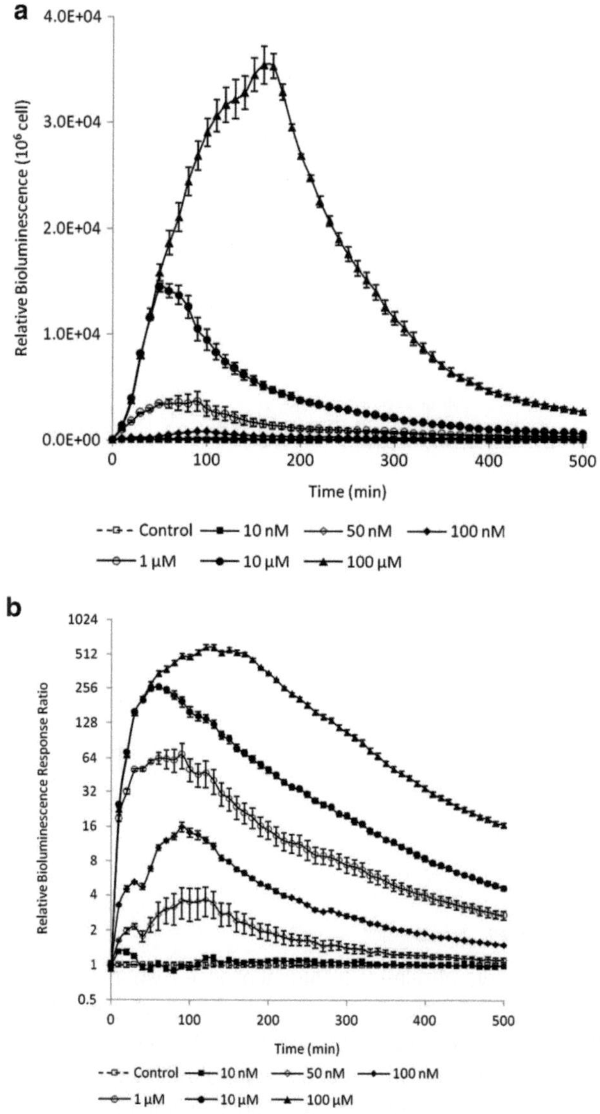

Fig. 1 Dynamic behavior of *Acinetobacter* ADPWH_lux (**a**) and bioluminescence response ratio (**b**) when induced by different concentrations of salicylate acid. (**a**) Relative bioluminescence of ADPWH_lux is shown as bioluminescent data per 10^6 cells. (**b**) Bioluminescence response ratios increase quickly only 10 min after addition of the inducer and reach a maximum at about 60–150 min

crude oil, and MMC are 100 μM, 600 μM, 0.1 mg/L, and 100 nM, respectively. The dynamics of the bioluminescence response ratio in Fig. 3a shows that each biosensor specifically responds to its relevant pollutant, which means in multi-contaminated samples, there is both a qualitative and a quantitative response allowing the biosensor to recognize and determine different specific pollutants concentrations, as shown in Fig. 3b.

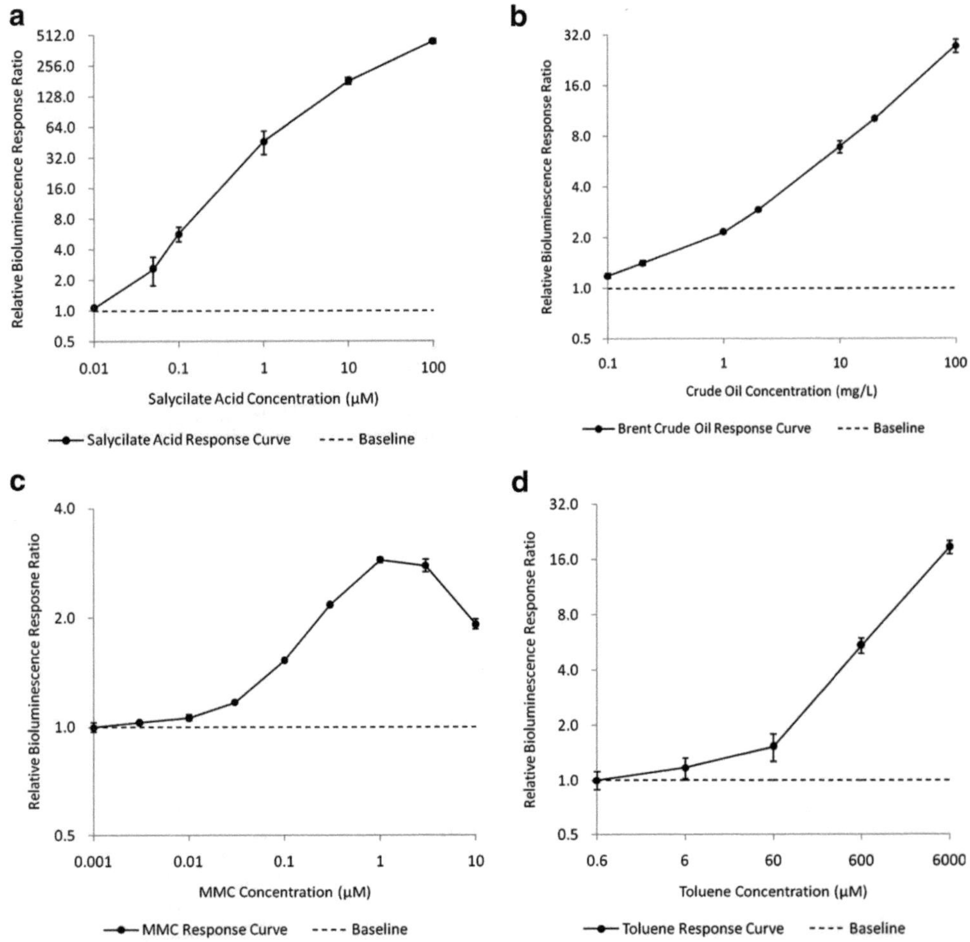

Fig. 2 Calibration curve of relative bioluminescence response ratio in pure water for specific biosensors. (**a**) Detection limit of ADPWH_lux for salicylate acid is 50 nM, and the effective response range is from 50 nM to 100 μM. (**b**) Detection limit of ADPWH_alk for Brent crude oil is 0.1 mg/L, and the effective response range is from 0.2 mg/L to 100 mg/L. (**c**) Detection limit of ADPWH_recA for mitomycin C is 3 nM, and the effective response range is from 3 nM to 3 μM. (**d**) Detection limit of ADPWH_Tol for toluene is 6 μM, and the effective response range is from 60 μM to 6 mM

A real oil-contaminated water sample was taken from sewage from a metal processing factory in Zhejiang Province, China, which was established as an automobile part manufacturing facility in 1996. A site investigation indicated that nearby groundwater and soil had been contaminated by total petroleum hydrocarbon (TPH) since the time the site was set up, mainly coming from the sewage, which received cutting liquid from the manufacturing workshops and contained mainly mineral oil, surfactant, and oil–water emulsion. The chemotoxicity of the original sewage was too high for biosensor detection (Fig. 4a, b), and therefore a series of different dilution factors (DF), namely 1, 5, 10, 50, 100, 500, and 1,000, were applied to find the optimal concentration for

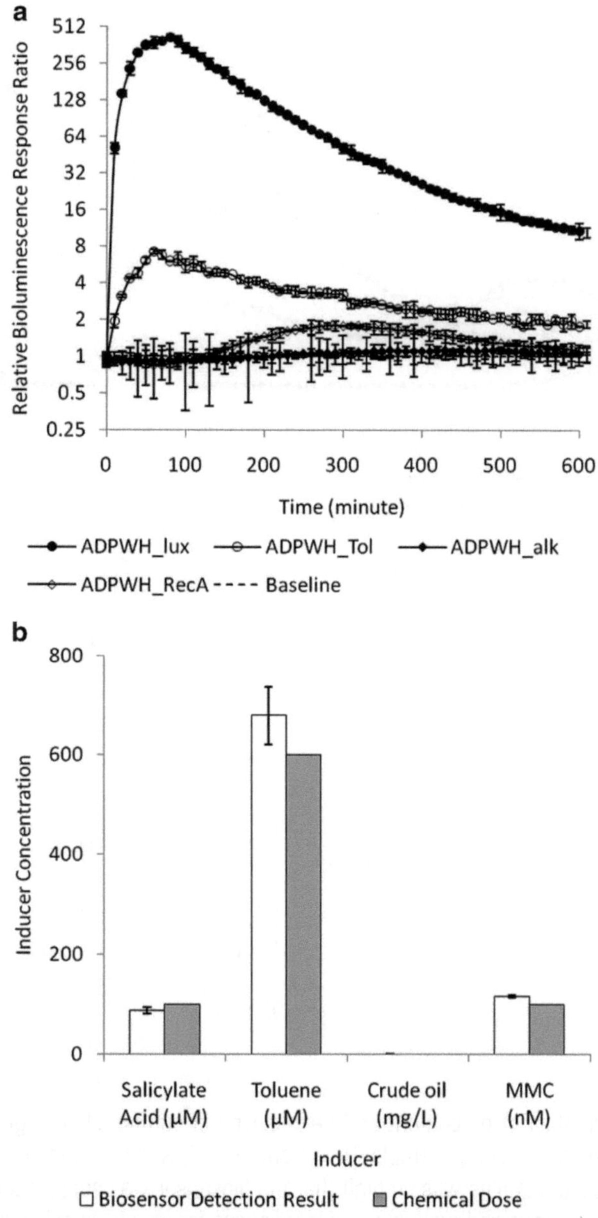

Fig. 3 Dynamic curve (**a**) and biosensor detection result (**b**) from an artificially polluted water sample. (**a**) ADPWH_lux, ADPWH_Tol, and ADPWH_recA show a significant response in the presence of inducer, while ADPWH_alk shows no response in the absence of crude oil. (**b**) Pollutant concentrations are calculated from the related calibration curves and match the theoretical dose, showing that whole-cell biosensor detection is both accurate and can be used to detect multi-contaminants in environmental samples

biosensor bioluminescent induction. The results suggested that the appropriate DF for ADPWH_alk and ADPWH_recA were 50–500 and 10–100, respectively. Other bioluminescence

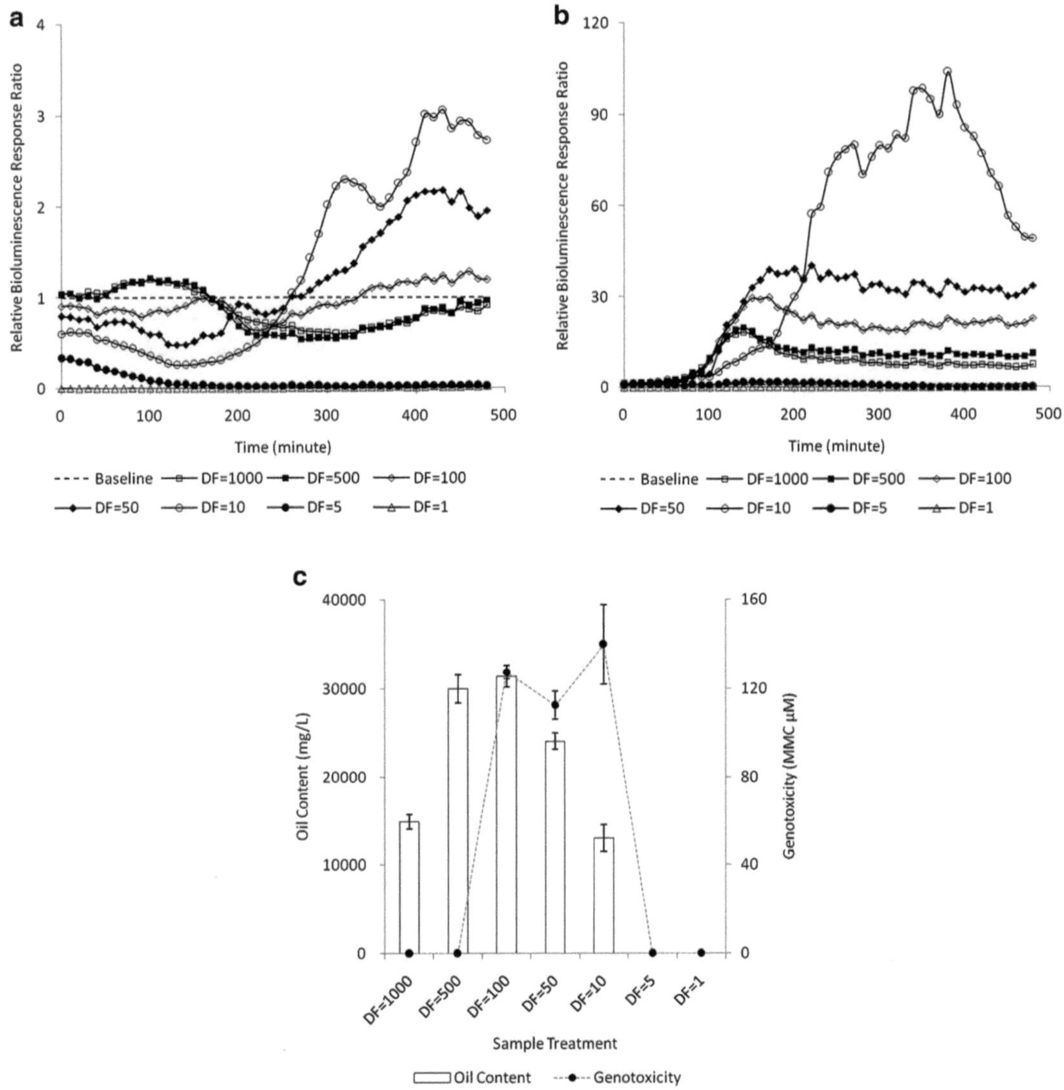

Fig. 4 Evaluation of oil content and genotoxicity in cutting oil sewage. (**a**) Bioluminescence of ADPWH_recA using the original sewage sample (DF = 1 and DF = 5) was below the baseline, indicating that the chemotoxicity of interrelated samples was so high that the biosensor was suppressed. The bioluminescence response ratio showed an obvious and gradual positive response for genotoxicity in suitably diluted samples (DF = 10, 50, and 100). (**b**) The appropriate dilution factor for ADPWH_alk detection was 50, 100, and 500, while the nonresponse and delay in expression at lower DFs was also caused by chemotoxic suppression. (**c**) Oil content and genotoxicity calculation, using an appropriate DF of 50–500 for ADPWH_alk and 10–100 for ADPWH_recA

responses suffer from chemotoxic suppression at lower DFs and become undetectable when the DF is too high.

A comparison with the calibration curve of ADPWH_alk and ADPWH_recA gave the oil content and genotoxicity of the sewage as shown in Fig. 4c. The genotoxicity of the sewage sample was equal to a MMC concentration of 126 μM, which represents a high risk to human health and the environment, but which is undetectable using a conventional analysis method. The oil content was 28,766 mg/L as detected using ADPWH_alk but was shown to be underestimated at 8,824 mg/L using a gravimetric method. This disparity was apparently caused by the pretreatment involved with the gravimetric method, which needed dichloromethane extraction and separation. As a result of the large amount of surfactant existing in sewage, an excessive oil–water emulsion was caused, leading to deficient dichloromethane extraction of the oil during the gravimetric pretreatment.

4 Notes

1. Needs to be autoclaved at 121 °C for 20 min.
2. Pass through a 0.20 μm disposable sterile filter to remove bacteria.
3. Biosensor concentration is about 5×10^9 CFU/mL in 1:1 diluted glycerin.
4. Best overnight culture time is from 12 h to 18 h when single colonies can be found on an agar plate or the cell concentration is about 10^8 CFU/mL in liquid medium.
5. Alternative centrifugation rates and times do not affect biosensor activity, e.g., 1,500 × *g* for 5 min or 3,500 × *g* for 3 min. Do not use a higher centrifugation rate than 3,500 × *g* to avoid cell death.
6. This step should be carried out between 0 and 4 °C.
7. *Acinetobacter* biosensors can remain active for 45 days at 4 °C and be ready for use if the above pretreatment is followed [13].
8. In real contaminated water or soil samples, pollutants are often found adsorbed to particles or located in porous media such as micropores or fissure interfaces which makes them more stable, harder to dissolve in water, and less bioavailable to bacteria making it more difficult for them to be taken up by the biosensor cells to trigger expression of the *luxCDABE* genes. Ultrasound is considered a good way to pretreat water or soil samples to emulsify pollutants into a water solution. The result of a series of exposure times (0, 30, 60, 120, 300, 600, 900, and 1,200 s) to 40 kHz ultrasound showed that the release equation obeyed the Korsmeyer–Peppas model, which represents the amount of drug released in a slow release system [15]. When diffusion is the main release mechanism, the square root of time will give a straight line against amount released, and under our experimental situations, the release mechanism deviates from the Fick equation, following an anomalous behavior (non-Fickian). Longer ultrasonic exposure time led to more pollutants being dissolved and an

Table 1
Recommended pollutant concentration ladders

Pollutant	Ladder (from highest to lowest concentration)						
Crude oil	100 mg/L	20 mg/L	10 mg/L	2 mg/L	1 mg/L	0.2 mg/L	0.1 mg/L
MMC	10 μM	5 μM	1 μM	100 nM	10 nM	5 nM	1 nM
Toluene	6 mM	2 mM	600 μM	200 μM	60 μM	20 μM	6 μM
Sodium salicylate	1 mM	500 μM	100 μM	10 μM	1 μM	100 nM	50 nM

exponential increase in bioluminescent response time. 300 s is proposed as an optimized time for water and soil sample pretreatment.

9. The photon emission spectra of the LuxCDABE protein extend from a wavelength of 400 nm to 700 nm [16], but the emitted bioluminescence will be absorbed by any suspended particles in the water samples or from the soil samples. High centrifugation rates can help to eliminate such absorption effects on bioluminescence and obtain a reproducible luminescence response.
10. Different volumes/weights are needed for the preparation of a standard storage solution of specific pollutants. Typically, 2.5 μL Brent crude oil, 668.8 mg MMC, 2.5 μL toluene, and 320.2 mg sodium salicylate are suggested to obtain final concentrations of 100 mg/L, 100 mM, 6.0 mM, and 100 mM, respectively.
11. Recommended concentration ladders for different pollutants are shown in Table 1.
12. It is possible to take more than four subsamples over a range of 5–200 mg depending on calibration requirement.
13. Ten minutes of preincubation at 30 °C will encourage biosensor physiological activities to give a better response during induction.
14. Though white microplates are most commonly used in luminescence detection to maximize the signal due to more light reflection, 96-well black clear bottom microplates are recommended here for biosensor bioluminescence detection. We found that interference between adjacent wells on the white plates was up to 3–5 %, while that between wells on black plates remained below 2 %. The background counts from white plates was always higher with huge variation, and fluorescence from the polyvinyl chloride material they are made from can also reduce detection sensitivity.

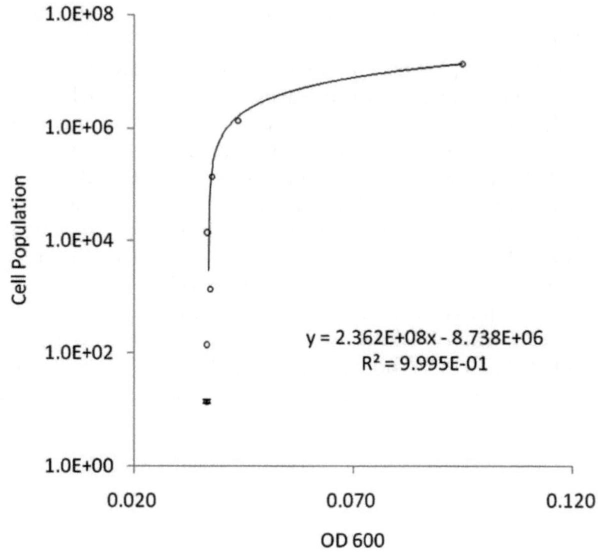

Fig. 5 Calibration curves of OD_{600} and cell population of *Acinetobacter* cells. Cell count-OD_{600} equation is cell count $= 2.362 \times 10^8 \times OD_{600} - 8.738 \times 10^6$

15. Cyclical shaking and measurement taking will influence cell growth and bioluminescence. A higher shaking frequency and intensity will help substrate diffusion and give a more homogeneous mix that can be taken up by the bacteria.
16. The cell population shows a strong linear relationship with optical density at 600 nm, though this can vary slightly between different microplate/plate readers (*see* Fig. 5).
17. Each biosensor has its own characteristic response peak time. Typically, ADPWH_lux always shows a rapid increase and reaches its maximal response at around 60–150 min and decreases quickly to form a sharp peak, while the bioluminescence of ADPWH_alk rises slowly, levelling off at its maximum expression at 150–250 min. The curve of ADPWH_Tol usually reaches its maximum at 180 min.

References

1. Rogers KR (2006) Recent advances in biosensor techniques for environmental monitoring. Anal Chim Acta 568:222–231
2. Vollmer AC, Van Dyk TK (2004) Stress responsive bacteria: biosensors as environmental monitors. Adv Microb Physiol 49:131–174
3. Rodriguez-Mozaz S et al (2005) Biosensors for environmental monitoring—a global perspective. Talanta 65:291–297
4. Belkin S (2003) Microbial whole-cell sensing systems of environmental pollutants. Curr Opin Microbiol 6:206–212
5. Rodriguez-Mozaz S, de Alda MJL, Barcelo D (2006) Biosensors as useful tools for environmental analysis and monitoring. Anal Bioanal Chem 386:1025–1041
6. DeFraia CT, Schmelz EA, Mou ZL (2008) A rapid biosensor-based method for quantification of free and glucose-conjugated salicylic acid. Plant Methods 4:28–38
7. D'Souza SF (2001) Microbial biosensors. Biosens Bioelectron 16:337–353
8. Mulchandani A, Bassi AS (1995) Principles and applications of biosensors for bioprocess

monitoring and control. Crit Rev Biotechnol 15:105–124
9. Huang WE et al (2006) Quantitative in situ assay of salicylic acid in tobacco leaves using a genetically modified biosensor strain of *Acinetobacter sp* ADP1. Plant J 46:1073–1083
10. Huang WE et al (2008) Characterizing the regulation of the Pu promoter in *Acinetobacter baylyi* ADP1. Environ Microbiol 10: 1668–1680
11. Zhang D et al (2011) Functionalization of whole-cell bacterial reporters with magnetic nanoparticles. Microb Biotechnol 4:89–97
12. Zhang D et al (2011) Whole cell bacterial bioreporter for actively searching and sensing of alkanes and oil spills. Microb Biotechnol 5:87–97
13. Song YZ et al (2009) Optimization of bacterial whole cell bioreporters for toxicity assay of environmental samples. Environ Sci Technol 43:7931–7938
14. Bauchop T, Elsden SR (1960) The growth of micro-organisms in relation to their energy supply. J Gen Microbiol 23:457–469
15. Costa P, Manuel J, Lobo S (2001) Modeling and comparison of dissolution profiles. Eur J Pharm Sci 13:123–133
16. Meighen EA (1991) Molecular-biology of bacterial bioluminescence. Microbiol Revs 55:123–142

Chapter 14

Stable Isotope Probing to Study Functional Components of Complex Microbial Ecosystems

Sophie Mazard and Hendrik Schäfer

Abstract

This protocol presents a method of dissecting the DNA or RNA of key organisms involved in a specific biochemical process within a complex ecosystem. Stable isotope probing (SIP) allows the labelling and separation of nucleic acids from community members that are involved in important biochemical transformations, yet are often not the most numerically abundant members of a community. This pure culture-independent technique circumvents limitations of traditional microbial isolation techniques or data mining from large-scale whole-community metagenomic studies to tease out the identities and genomic repertoires of microorganisms participating in biological nutrient cycles. SIP experiments can be applied to virtually any ecosystem and biochemical pathway under investigation provided a suitable stable isotope substrate is available. This versatile methodology allows a wide range of analyses to be performed, from fatty-acid analyses, community structure and ecology studies, and targeted metagenomics involving nucleic acid sequencing. SIP experiments provide an effective alternative to large-scale whole-community metagenomic studies by specifically targeting the organisms or biochemical transformations of interest, thereby reducing the sequencing effort and time-consuming bioinformatics analyses of large datasets.

Key words Stable isotope probing, Ecosystem functioning, Microbial ecology, Functional metagenomics

1 Introduction

Microbial ecology studies have traditionally relied on bacterial isolation and cultivation techniques to understand the diversity and infer the activity of environmental microorganisms. However, this approach is severely limited by the availability of suitable isolates and the recognition that the vast majority of environmental species have resisted efforts to cultivate them in the laboratory [1]. Culture-based studies also suffer from being reductionist in approach. Biochemical processes in nature are frequently carried out in the context of a network of microorganisms, or cooperative consortia, and rarely by a single species in isolation. Hence, cultivation studies with biochemical analyses of pure isolates can present an incomplete picture of the chemical transformations undertaken by microbial communities in situ.

Ian T. Paulsen and Andrew J. Holmes (eds.), *Environmental Microbiology: Methods and Protocols*, Methods in Molecular Biology, vol. 1096, DOI 10.1007/978-1-62703-712-9_14, © Springer Science+Business Media, LLC 2014

Molecular methods have been developed to provide insights into the diversity of microbial communities inhabiting natural environments without the need to cultivate them in the laboratory. Methods such as the polymerase chain reaction (PCR) and fluorescence in situ hybridization (FISH) targeting the 16S rRNA gene have documented the enormous diversity of environmental microbes. More recently, random or "shotgun" sequencing of community DNA extracted directly from a particular environment has been used to reconstruct the metabolic potential of a microbial community by piecing together the genes and genomes of its members (e.g., [2]). While this approach, known as metagenomics, has successfully described relatively simple systems, such as acid mine drain communities [2], the ability to reconstruct the metabolic potential of a natural habitat decreases dramatically as the complexity of the microbial community in the sample increases. For most natural seawater and soil environments, the amount of sequence coverage required is sufficiently large that it is near impossible to reconstruct the entire community. As a consequence, the DNA of keystone species, i.e., those that make a significant contribution to ecosystem function despite low abundance, can easily be overwhelmed by the DNA of a few dominant species, and therefore important ecosystem functions can be completely missed.

Stable isotope probing (SIP) offers the ability to study organisms involved in a biochemical transformation of interest by targeting the species capable of using a specific growth substrate [3, 4]. SIP does not involve the use of radioisotopes and is therefore a portable and versatile technique that can be applied to a wide range of environments and conditions. SIP has been shown to be successful with both ^{13}C- and ^{15}N-labelled compounds (see example studies summarized in Table 1), which are incorporated into the

Table 1
DNA stable isotope probing has been applied on a wide range of substrate and environments

SIP substrate	Environment	Targeted microbial group	Reference
^{13}C methanol, methane	Acidic forest soil	Methylotrophic bacteria	[5]
^{13}C methanol, methylamine	Surface coastal seawater	Microbial one-carbon degraders	[11]
^{13}C carbon dioxide	Biofilm	Autotrophic ammonia-oxidizing bacteria	[12]
^{13}C glucose, lactate	Supragingival plaque	Bacteria involved in oral caries onset	[13]
^{13}C glucose, phenol, caffeine, naphthalene	Soil	Global microbial community	[14]
^{15}N ammonium sulfate	Pure cultures and Westola soil	Global microbial community	[15]
^{15}N dinitrogen	Uncultivated soil	Diazotrophs	[16]
^{15}N-(ring) hexahydro-1,3,5-trinitro-1,3,5-triazine	Aquifer sediments and groundwater	Global microbial community	[17]

nucleic acid pool of the active microbes and differentiate them against the other community members who do not utilize the substrate. SIP has been applied to genomic (DNA-SIP) [5] and transcriptomic studies (RNA-SIP) [6], cell lipids (PLFA-SIP) [7], and proteins (protein-SIP) [8]. A number of detailed protocols for DNA-SIP have been described previously [9, 10], and we refer the readers to these publications. In this chapter, we will emphasize the versatility of stable isotope probing in environmental microbiology and highlight the advantages of this approach for the study of prokaryotic functional diversity in ecology studies.

2 Materials

Prepare all solutions using analytical grade reagents and ultrapure water (pre-filtered through 0.2 μm pore-size filters and deionized to attain a sensitivity of 18.2 MΩ cm at 25 °C). Ideally, equipment for environmental sampling should be sterilized, but this is not always practical when working in the field. As an alternative, before sampling, equipment can be washed with hot water (>60 °C) and neutral detergent (e.g., 1.0 % v/v neutracon, Decon Laboratories Ltd, Hove, UK) and then rinsed thoroughly with hot water to remove traces of detergent before a final rinse with Milli-Q water (Millipore Ltd, Watford, UK).

2.1 Environmental Sampling

1. Sampling apparatus adapted to the environment, e.g., acrylic sediment corers/tubes for sediments and soils and bottles/ buckets for water.
2. Data acquisition apparatus, e.g., thermometer, oxygen probe, and light meter.
3. Spatulas, analytical balance, sterile water, and tissue paper.
4. Thermo-control carrier for transport of the samples from field to laboratory in temperature stable conditions.
5. Ethanol 70 %.

2.2 Stable Isotope Probing Experiment

1. Spatula, analytical balance.
2. Eppendorf tubes 1.5 mL.
3. Plastic tubes 50 and 15 mL.
4. Set of pipettors to dispense volumes from 1 to 1,000 μL, with sterile filter tips.
5. ^{13}C- or ^{15}N-labelled compound, sufficient amount for replicates and time series.
6. Incubation vials, e.g., 125 mL Wheaton glass serum bottles with crimp top (Sigma).
7. Septum seals and crimps (for gaseous substrates).
8. Temperature-controlled incubation chamber (or adequate in situ location).

9. Adequate detector for the substrate used (e.g., gas or liquid chromatography, chemical assay).
10. Ice/liquid nitrogen.

2.3 Nucleic Acid Extraction and Cesium Chloride Gradient Separation

1. Clean sterile environment to prepare the nucleic acid and the cesium chloride gradient, e.g., UV laminar flow.
2. Set of pipettors to dispense volumes from 1 to 1,000 μL, with sterile DNA-/RNA-free filter tips.
3. Nucleic acid extraction method (e.g., FastDNA SPIN Kit for soil, MP Bio).
4. Nuclease-free sterile water (e.g., nuclease-free water, Ambion®, UK).
5. Cesium chloride solution: 7.163 M or 1.890 g/L in Milli-Q water.
6. Gradient buffer (GB): 0.1 M Tris–HCl, 0.1 M KCl, 1 mM EDTA, pH 8.0.
7. PEG DNA precipitation solution: 30 % polyethylene glycol 6000, 1.6 M NaCl.
8. Glycogen solution: 20 mg/mL in water (e.g., UltraPure Glycogen, Invitrogen).
9. TE buffer: 10 mM Tris–HCl, 1 mM EDTA, pH 8.0.
10. Refractometer (e.g., Reichert, AR2000 digital) (*see* **Note 1**).
11. Ultracentrifuge (e.g., XL80, Beckman Coulter), rotor (e.g., VTi 65.2, Beckman Coulter), and tubes and tube heat sealer (e.g., 5.1 mL polyallomer Quick-Seal® ultracentrifugation tube for vertical centrifugation, Beckman Coulter).
12. Fractionation setup: stand and clamp, needles, tubing, peristaltic pump and recipient tubes.
13. Microcentrifuge, speed capacity up to 10,000 ×*g*.
14. Quant-iT™ PicoGreen® dsDNA assay kit (Invitrogen, Carlsbad, CA, USA). Store at 4 °C.
15. Fluorescence microplate reader with the following filters: excitation ~480 nm (i.e., fluorescein) and emission ~520 nm (e.g., 485 nm/em.; 535 nm is suitable).

3 Methods

The method below is a guide for the setup and implementation of DNA-SIP experiment, with particular attention given to the major pitfalls and considerations for experimental design. The method is not intended as a descriptive step-by-step protocol. For this, readers are directed to previous publications [9, 10].

3.1 Collection of Environmental Samples

Stable isotope probing experiments rely on the activity of a portion of the bacterial communities present in the samples. Hence, some practical considerations need to be addressed when carrying out the sample collection, storage, and transport in order to minimize bias in the following analysis. It is important to assess the environmental conditions relating to the sample studied. For each collection, the environmental metadata, such as temperature, oxygen status (e.g., for sediments), humidity (particularly important for soil samples), and light (e.g., for the euphotic zone of aquatic samples), should be recorded. These physical and chemical parameters will be used to set up the SIP incubations (*see* Subheading 3.2).

During sampling, and in particular for highly structured sediment or soil cores, the collection needs to be conducted carefully to avoid cross contamination of functionally distinct strata which could introduce a sampling bias that would be amplified during the incubation step. Clean utensils and equipment should always be used. Even though it is often impossible to maintain sterile equipment during field sampling campaigns, it is possible to rinse and clean the utensils in-between samples, using sterile Milli-Q water and ethanol (70 % v/v).

Depending on the conditions and rate of activity of the microbial community (*see* **Note 2**), stable isotope probing (SIP) experiments can be initiated at the collection site with proper preparation (*see* **Note 3**). Alternatively, the collected sample can be brought back into the laboratory (*see* **Note 4**). Either way, to obtain the best results from a SIP incubation, it is important to maintain the physical and chemical conditions of the samples and to minimize variation of temperature while limiting the time before the SIP incubation is started. This is of special importance when conducting RNA-SIP experiments, where time of incubation is shorter and less incorporation of substrate is needed.

The main considerations for proper sampling are:

1. Recording of in situ parameters (physical and chemical).
2. Proper containers and sampling procedures.
3. Determine the quantity of sample required for sufficient replicate and control incubations.
4. Adequate transport conditions of samples.

3.2 Stable Isotope Probing Experiment: Incubation Time Series

It is preferable to perform preliminary substrate uptake experiments using a range of substrate concentrations in order to verify the biological activity of the microbial populations from the selected site (*see* **Note 5**). This will also help to establish the incubation time required to ensure sufficient incorporation of stable carbon or nitrogen atoms in the active microbes. A minimum of 5 μmol of labelled carbon per gram of sample is required to ensure the adequate separation of the "heavy" nucleic acid, i.e., the genetic material with incorporated stable isotope atoms such as ^{13}C or ^{15}N, from the "light" nucleic acid, with the more naturally abundant ^{12}C and ^{14}N atoms. However, if possible, the separation of heavy and light fractions will be improved if higher label incorporation levels can be achieved with higher label concentrations, e.g., 50 μmol carbon per gram of sample. It is markedly more important to use higher substrate concentrations for $^{15}N/^{14}N$ SIP experiments as fewer nitrogen atoms are incorporated per unit of nucleic acid resulting in a mass shift that is less pronounced.

In order to understand the uptake dynamics of the active members of a microbial community towards a specific compound, we recommend carrying out a time series incubation where an aliquot or replicate incubation is sampled at regular time intervals (or at a defined level of substrate incorporation into biological material, e.g., 15, 50, 100 μmol of C/N total). For example, during incubation experiments with ^{13}C-DMS (dimethylsulfide), we examined the disappearance of the substrate from the incubation headspace by gas chromatography mass spectrometry (GC-MS) over time. This allowed us to detect a range of biochemical activities, i.e., rapid and slow DMS consumers, presumably due to distinct populations present at the start of the experiment or a succession of organisms utilizing DMS.

The time and level of incorporation of the labelled compound with stable isotope C/N is, in part, specific to each substrate; thus, any SIP experiment will need to be optimized. The time and level of incorporation of the labelled substrate must be sufficient to obtain proper labelling and separation of the active microbes towards the compound but also needs to be short enough to ensure the heavy carbon or nitrogen atoms are only appearing in primary degraders of the substrate without transfer to secondary feeders, e.g., through metabolic network/transfer of the atoms from lysis of the cell, cellular death of the primary feeder, or utilization of metabolites produced.

Another significant aspect is the establishment of controls and references to account for bias or alterations to the bacterial community independently from the substrate uptake. Alongside the heavy isotope incubations, control incubations with "light" substrate (^{12}C or ^{14}N) need to be performed as comparative references. These will indicate alterations in community structure attributable to the incubation conditions. Some environmental samples can have a high intrinsic heterogeneity, e.g., chemical gradients in soil and sediments. Hence, it is crucial to set up biological replicates for any stable isotope studies, with both controls and labelled substrates. Furthermore, some aspects of incubation such as the supplementation of the labelled substrate can introduce a change in conditions, and it is vitally important to design appropriate controls that will define changes such as dilutions or temperature variation.

The establishment of SIP experiments for environments with very low microbial loading or very poor nutrient conditions can be a challenge as the uptake of substrate is likely to require lengthy incubation times. For such samples, it is worth considering the supplementation of nutrients or the concentration of the bacterial load, e.g., filtration of cells in suspension from very dilute freshwater systems.

The main considerations for a successful SIP experiment design are:

1. Conduct preliminary uptake and substrate detection tests, and prepare incubation setup, such as conditions, and material needed.
2. Conserve original samples as reference.
3. Perform time series of incorporation with heavy isotopes ($^{13}C/^{15}N$).
4. Carry out control incubations with light isotopes ($^{12}C/^{14}N$) mirroring the incubations with heavy isotopes.
5. Perform replicates for any incubation, with both heavy and light isotopes.

3.3 Nucleic Acid Extraction and Cesium Chloride Gradient Separation for DNA-SIP

The original protocol describing the SIP-DNA density gradient separation contained ethidium bromide (EtBr) to allow the visualization of the nucleic acid [9]. Since the establishment of the gradient is not influenced by the presence of EtBr, we prefer to perform a fractionation of the resulting gradient, which removes the need to visualize the DNA and limits operator exposure to high concentrations of EtBr and UV light [18]. This will also eliminate the deleterious effects of UV illumination on the nucleic acid extracted:

1. Extraction of nucleic acids from the SIP experiment samples, e.g., from salt marsh sediment samples (450–500 mg sediment) using the Fast Soil Extraction Kit (MP Bio) following the manufacturer's instruction, using a mechanical lysing step (6,000 rpm for 40 s in a tissue homogenizer, e.g., Precellys® 24 (Bertin Technologies, France)).
2. Quantification of the obtained nucleic acid is carried out in triplicate on tenfold dilutions of the extracted DNA (target concentration around 1–10 ng/μL) against a standard curve prepared from a DNA sample of known concentration (e.g., lambda DNA standard) using the dsDNA-specific fluorescent dye PicoGreen (*see* **Note 6**). Briefly, using black fluorescence 96-well microplates, a standard curve is produced from serial dilutions of a lambda DNA standard (provided in the PicoGreen assay kit) alongside the samples to be quantified (e.g., 1 μL of the diluted sample diluted in 99 μL of TE buffer) (*see* **Note 7**). The freshly prepared dye solution (100 μL

of 1× concentration; *see* **Note 8**) is mixed with each sample and incubated for 3–5 min at room temperature before analysis with a fluorescence microplate reader using with excitation ~480 nm and emission ~520 nm.

3. To retrieve sufficient nucleic acid from the density gradient separation, it is recommended to begin with a minimum of 3 μg of nucleic acid on the CsCl gradient (*see* **Note 9**).
4. Prepare the solution to establish a continuous cesium chloride (CsCl) gradient as described previously (*see* **Note 10**) [10]. Briefly, prepare 5.8 mL of a mix of the CsCl stock solution (1.886 g/mL) with gradient buffer (GB, 1 g/mL) to a final density of 1.725 g/mL in a sterile tube (e.g., 15 mL tube). Combine the nucleic acid with the prepared solution (CsCl/GB). Carefully fill an ultracentrifugation tube with the mix, until flush with the tube opening (*see* **Note 11**), and balance the tube with a pair to within 10 mg (*see* **Note 12**). If the experiment gives an odd number of samples/tubes for the centrifugation, prepare a balance blank gradient in the same way as the sample gradient tubes. Seal the tubes with a heat sealer and carefully verify that no leaks are present in any of the prepared tubes.
5. Place tube pairs in a vertical centrifugation rotor (e.g., VTi 65.2). Allow the gradient to form at 44,100 × *g* for a minimum of 36 h at 20 °C in vacuum. It is crucial to not apply the centrifuge brakes during the deceleration phase at the end of the centrifugation, as this would disturb the gradient (*see* **Note 13**).
6. Following the careful extraction of the tubes from the rotor (*see* **Note 14**), slowly fractionate the gradient by density using a peristaltic pump. This is achieved by carefully piercing the top of the tube with a syringe needle attached to small-bore platinum-cured silicone tubing (e.g., internal diameter of 2 mm) fed through a peristaltic pump (set at 0.5 mL/min) and then piercing a hole at the bottom of the gradient tube with another syringe needle (*see* **Note 15**). Fractions of about 0.5 mL are collected by forced gravity into recipient tubes placed below the gradient tube. Fractionation is made more consistent by replacing the extracted fraction with water containing dye such as bromophenol blue. A full gradient usually retrieves 12–13 fractions from the first "heavy" to the last "light."
7. It is recommended to verify the gradient formation by measuring the density of every fraction collected with a refractometer (*see* **Note 1**).
8. Precipitate the nucleic acid contained in the collected fractions. Add 2 μL glycogen solution (e.g., Invitrogen) to each fraction, and mix by inversion. Add approximately 2 volumes (800 μL) of PEG solution, and further mix by inversion. Leave the mixture to precipitate at room temperature for a minimum of 2 h,

and then pellet the precipitated nucleic acids by centrifugation at 10,000×*g* for 30 min at 4 °C. Discard the supernatant and wash the pellet with 70 % ethanol; centrifuge at 10,000×*g* for 10 min at 4 °C. Evaporate the ethanol and resuspend the nucleic acid pellet into 30 μL of nuclease-free water or adequate buffer (e.g., TE).

9. Quantification of the nucleic acid from the various fractions can be carried out as described in Subheading 3.3, **step 2**.

3.4 Applications of the Stable Isotope Probing Experiment

DNA-SIP has been applied to a wide range of samples with numerous labelled substrates. The stable isotope probing technique has been extended to analyze other components of active microbes in conjunction to the genomic DNA to obtain a more complete insight of specific biochemical cycles. Successful RNA fractionation (RNA-SIP), phospholipid fatty acid (PLFA-SIP), and protein (protein-SIP) studies have been carried out:

1. The study of phospholipid fatty acid (PLFA) by the use of stable isotope was actually reported before nucleic acid stable isotope probing. Boschker and colleagues [7] investigated aquatic sediment samples with ^{13}C acetate and methane. PLFA-SIP can be combined with other SIP techniques to offer a more complete picture of the microbial network participating to a particular biological cycle. PLFA-SIP has been successfully implemented alone [19] or in conjunction with DNA-SIP [20].
2. A methodology similar to the DNA-SIP technique was adapted to conduct targeted transcriptomics studies (RNA-SIP). Analysis can be carried out in parallel to DNA-SIP to give more insights into the response of active microbial population and the functional genes involved in particular nutrient cycles. For RNA-SIP, levels of incorporation of μmol of C or N needed are greatly reduced (from a tenth to a hundredth of the amount required for DNA-SIP). Furthermore, the stable isotope is assimilated faster into RNA molecules than in the DNA (from 24 to 48 h) [6]. This means that the required incubation times will be much shorter.
3. Stable isotope probing has been adapted to resolve the protein component of metabolically active species [8, 21]. Heavy isotope atoms are incorporated at higher levels into proteins than DNA/RNA making protein-SIP more sensitive than nucleic acid SIP. The mass difference calculated between unlabelled and labelled peptides yields quantitative data of the uptake and incorporation of the substrate into the active microbial cells. Protein-SIP provides both phylogenetic and functional information that can be quantitative and be carried out alongside DNA-SIP to obtain a wider genomic context.

The nucleic acid recovered from DNA-SIP experiments can be further used for an array of analyses such as:

1. Amplification of specific phylogenetic markers or functional genes, library construction and sequencing.
2. General microbial community "fingerprinting" from different fractions, e.g., DGGE and tRFLP (e.g., [11]).
3. Multiple displacement amplification (MDA) also known as "whole-genome amplification" of fractions of interest for shotgun sequencing or large-fragment library construction [22–25].

4 Notes

1. If no refractometer is available, the measurement of the density of the solutions can be carried out with an analytical balance and calibrated pipette. However, it is important to carefully and precisely measure the density of the cesium chloride stock solution for the successful establishment of the density gradient.
2. Before the establishment of the stable isotope probing experiment, it is preferable to estimate the activity towards the substrate used in the selected environment, as described in Subheading 3.2. This is sometimes impractical for field work, but preliminary results will inform the design and quality of the SIP experiment.
3. The SIP experiment can be carried out or initiated in situ. In both cases, it is important to ensure incubation conditions are kept consistent and do not fluctuate so as to alter the microbial community independently from the biochemical activity studies.
4. The integrity of the samples needs to be conserved during transport from the sampling site to the laboratory. SIP experiments rely on the biochemical activity of live microbial communities; hence, the metabolic enrichment must be initiated rapidly and will be most successful on freshly collected samples.
5. SIP experiments require the use of environmentally relevant conditions in relation to both the physical and chemical parameters measured from the site of sampling but also regarding the level of substrate to use. Some compounds can apply a selective pressure or become toxic at high concentration, and it might be needed to carry out sequential additions of labelled substrate to maintain close to in situ concentrations.
6. Determination with the usual absorbance at 260 nm, e.g., using a NanoDrop (Thermo Scientific), cannot be used due to the presence of saturating quantities of nucleotides and random priming hexamers in the reaction buffers. These interfere with the reading producing inaccurate concentration estimations.

7. Ensure that the solutions used to prepare the dilutions and blanks are prepared with nuclease- and nucleic acid-free water.
8. The PicoGreen fluorescent dye is light sensitive; ensure it is protected from light by wrapping tubes with foil or using light excluding microplate covers. The fluorescent dye is conserved in DMSO, which becomes solid at 4 °C. The stock solution of dye should be thawed at room temperature for about 20 min before use to ensure the whole volume of dye solution is dissolved before diluting in TE buffer. Do not overheat the solution to accelerate the process as it may damage the dye stock. Once developed, the fluorescence signal remains stable for a few hours at room temperature.
9. The separation and fractionation of the DNA recovered from the SIP experiment can be conducted with less than the recommended amount of at least 3 μg; however, there will be a low recovery of "heavy" labelled DNA. It is recommended to increase the initial quantity of sample processed in the SIP experiment if insufficient nucleic acids can be extracted.
10. To obtain uniform and comparable continuous gradients between the samples and references of a SIP experiment, it is best to prepare the working CsCl solution from the same CsCl stock solution. Even small variations in the solution density can greatly influence the proper formation of the continuous gradient.
11. The ultracentrifugation tubes have to be fully filled flush to the top opening, and proper sealing has to be ensured. If a bubble of air is left or the tube is improperly sealed, there is a risk of the tube bursting during high-speed centrifugation.
12. The density gradients are formed at high speeds of centrifugation. Small imbalances can prove dangerous mechanically but also impose strain on the machine, leading to the risk of mechanical failure. Furthermore, if the tubes are not properly balanced, it will alter the gradient formation, and the separation between labelled and unlabelled nucleic acids will be poor.
13. The deceleration without brake will approximately add an extra 2 h to the overall run time.
14. Care must be taken during the extraction of the tubes after centrifugation and during the fractionation. The density gradient established in the tube can be disturbed easily, and the various separated labelled and unlabelled nucleic acid layers could be mixed.
15. The pressure applied by the manipulator must be carefully measured when piercing the gradient tube with the syringe needles to ensure that the needle does not disturb the gradient layers in the tube.

Acknowledgements

Work in the laboratory of HS is supported by the NERC.

References

1. Rappe M, Giovannoni SJ (2003) The uncultured microbial majority. Annu Rev Microbiol 57:369–394
2. Tyson GW et al (2004) Community structure and metabolism through reconstruction of microbial genomes from the environment. Nature 428:37–43
3. Dumont MG, Murrell JC (2005) Stable isotope probing—linking microbial identity to function. Nat Rev Microbiol 3:499–504
4. Friedrich MW (2006) Stable-isotope probing of DNA: insights into the function of uncultivated microorganisms from isotopically labeled metagenomes. Curr Opin Biotechnol 17:59–66
5. Radajewski S et al (2002) Identification of active methylotroph populations in an acidic forest soil by stable-isotope probing. Microbiology 148:2331–2342
6. Manefield M et al (2002) RNA stable isotope probing, a novel means of linking microbial community function to phylogeny. Appl Environ Microbiol 68:5367–5373
7. Boschker HTS et al (1998) Direct linking of microbial populations to specific biogeochemical processes by C-13-labelling of biomarkers. Nature 392:801–805
8. Jehmlich N et al (2008) Protein-based stable isotope probing (protein-SIP) reveals active species within anoxic mixed cultures. ISME J 2:1122–1133
9. Radajewski S et al (2000) Stable-isotope probing as a tool in microbial ecology. Nature 403:646–649
10. Neufeld JD et al (2007) DNA stable-isotope probing. Nat Protoc 2:860–866
11. Neufeld JD et al (2007) Stable-isotope probing implicates Methylophaga spp and novel Gammaproteobacteria in marine methanol and methylamine metabolism. ISME J 1:480–491
12. Avrahami S et al (2011) Active autotrophic ammonia-oxidizing bacteria in biofilm enrichments from simulated creek ecosystems at two ammonium concentrations respond to temperature manipulation. Appl Environ Microbiol 77:7329–7330
13. McLean JS et al (2012) Identifying low pH active and lactate-utilizing taxa within oral microbiome communities from healthy children using stable isotope probing techniques. PLoS One 7:e32219
14. Padmanabhan P et al (2003) Respiration of ^{13}C-labeled substrates added to soil in the field and subsequent 16S rRNA gene analysis of ^{13}C-labeled soil DNA. Appl Environ Microbiol 69:1614–1622
15. Cupples AM et al (2007) DNA buoyant density shifts during ^{15}N-DNA stable isotope probing. Microbiol Res 162:328–334
16. Buckley DH et al (2007) Stable isotope probing with $^{15}N_2$ reveals novel noncultivated diazotrophs in soil. Appl Environ Microbiol 73:3196
17. Roh H et al (2009) Identification of hexahydro-1,3,5-trinitro-1,3,5-triazine-degrading microorganisms via 15N-stable isotope probing. Environ Sci Technol 43:2505–2511
18. Lueders T, Manefield M, Friedrich MW (2004) Enhanced sensitivity of DNA- and rRNA-based stable isotope probing by fractionation and quantitative analysis of isopycnic centrifugation gradients. Environ Microbiol 6:73–78
19. Evershed RP et al (2006) ^{13}C-Labelling of lipids to investigate microbial communities in the environment. Curr Opin Biotechnol 17:72–82
20. Webster G et al (2006) A comparison of stable-isotope probing of DNA and phospholipid fatty acids to study prokaryotic functional diversity in sulfate-reducing marine sediment enrichment slurries. Environ Microbiol 8:1575–1589
21. Jehmlich N et al (2010) Protein-based stable isotope probing. Nat Protoc 5:1957–1966
22. Chen Y et al (2008) Revealing the uncultivated majority: combining DNA stable-isotope probing, multiple displacement amplification and metagenomic analyses of uncultivated Methylocystis in acidic peatlands. Environ Microbiol 10:2609–2622
23. Chen Y, Murrell JC (2008) DNA-stable isotope probing: recent advances and protocols. In: Rapley R (ed) Aquino de Muro, M. Methods in Molecular Biology—Gene Probes. Humana Press, Totowa, NJ
24. Chen Y et al (2010) Metagenomic Analysis of Isotopically Enriched DNA. In: Streit WR, Daniel R (eds) Metagenomics: Methods and Protocols, vol 668, Methods in Molecular Biology, vol., pp 67–75
25. Chen Y, Murrell JC (2010) When metagenomics meets stable-isotope probing: progress and perspectives. Trends Microbiol 18:157–163

Part IV

High Throughput Genomic Approaches for Environmental Microbiology

Chapter 15

Metagenomics Using Next-Generation Sequencing

Lauren Bragg and Gene W. Tyson

Abstract

Traditionally, microbial genome sequencing has been restricted to the small number of species that can be grown in pure culture [1]. The progressive development of culture-independent methods over the last 15 years now allows researchers to sequence microbial communities directly from environmental samples. This approach is commonly referred to as "metagenomics" or "community genomics". However, the term metagenomics is applied liberally in the literature to describe any culture-independent analysis of microbial communities. Here, we define metagenomics as shotgun ("random") sequencing of the genomic DNA of a sample taken directly from the environment. The metagenome can be thought of as a sampling of the collective genome of the microbial community. We outline the considerations and analyses that should be undertaken to ensure the success of a metagenomic sequencing project, including the choice of sequencing platform and methods for assembly, binning, annotation, and comparative analysis.

Key words Metagenomics, Microbial ecology, Roche 454, Illumina

1 Metagenomics: Advancing the Field of Microbial Ecology

Metagenomics is a relatively new addition to the molecular toolbox for microbial ecologists and is the most direct, unbiased means to interrogate the functional potential of microbial communities. The first large-scale metagenomic sequencing study performed shotgun sequencing of two viral communities found in surface seawater [2]. Over 65 % of the viral sequences found in this study were novel, with the dominant population responsible for only 2–3 % of the sequences generated. In 2004, one of the first microbial metagenomic projects was performed by Venter and colleagues to examine microbial populations within surface waters of the Sargasso Sea [3]. They estimated that over 1,800 genomic species were in the sample, including 148 novel bacterial phylotypes [3]. Despite generating vast (1.045 Gb) amounts of Sanger sequence data, only 25 % of the reads could be assembled. While the pioneering metagenomic studies were conducted using the Sanger platform [2–9], it was quickly recognized that this technology could not provide sufficient

Ian T. Paulsen and Andrew J. Holmes (eds.), *Environmental Microbiology: Methods and Protocols*, Methods in Molecular Biology, vol. 1096, DOI 10.1007/978-1-62703-712-9_15, © Springer Science+Business Media, LLC 2014

read depth to saturate moderately diverse communities. However, partial and near-completed genomes have been recovered using Sanger sequencing of relatively simple communities [9]. In the absence of adequate sequencing depth for metagenome assembly, the use of gene-centric analyses can yield valuable biological findings [10]. Examples include the identification of novel biomass-degrading genes in cow rumen [11] and the observation of functional selection in communities as a consequence of extreme artificial perturbation [12]. These studies use sequence clustering as opposed to assembly to reduce dataset complexity.

The clear benefits of next-generation sequencing have promoted a substantial increase in the number of metagenomic studies undertaken, with over 300 studies in progress, or completed (GOLD [13]). These studies span a wide spectrum of environments, including clinical [6, 14], engineered [15], and natural communities [16–19]. Of next-generation platforms, there was initially a clear preference towards the long reads produced by Roche 454 pyrosequencing [15, 19–22]. However, substantial improvements in Illumina throughput and read length have seen this platform increase in popularity for metagenomic studies [23–25].

Despite the throughput of next-generation platforms, sequencing should not be blindly applied to an environmental sample without devising an appropriate sequencing strategy. The strategy used (single or multiple platforms, fragment or paired libraries) must take into account both the research question and the composition of the target community. Failure to do so can severely hamper downstream processes and the overall success of the metagenomic study.

2 Understanding How Underlying Community Composition Influences Metagenomics

In metagenomics, the DNA is extracted directly from an environmental sample. Factors such as the genome size, genome copy number, within-species heterogeneity, and the relative abundance of the species, as well as biases in DNA extraction and sequencing, will determine the number of reads that are derived from a given species. Shotgun sequencing of the community typically results in deep sequencing of the numerically dominant species, with very few reads from the rarer community members. Thus, the amount of sequencing required to saturate the community is positively correlated with the community diversity and population complexity. Research investigating the functional capacity of "rare" species (<1 % relative abundance) may find artificially increasing the abundance of the target species (using either enrichment or flow cell-sorting methods) more tractable than the cost of generating and handling the amount of sequencing data required to capture this functionality from the metagenome.

The use of approaches complimentary to metagenomics, such as directed sequencing of the 16S rRNA gene amplicons, can help determine the choice of sequencing platform(s), library types, and amount of sequencing required to obtain the datasets necessary to achieve research goals.

3 Sequencing Platforms for Metagenomics

In recent years, a number of new sequencing technologies have emerged that increase the feasibility of metagenomic projects. These newer technologies provide cheaper, faster, and higher-throughput sequencing. The diversity and non-uniform abundance of microbial communities make high-throughput sequencing essential for obtaining adequate coverage of community members. With each sequencing technology having distinct benefits and drawbacks, the choice of technology used can ultimately determine the relative success of a metagenomic project. Here, the pros and cons of the most widely used platforms for metagenomics will be discussed.

3.1 Sanger Sequencing

Chain-terminator sequencing (more frequently referred to as "Sanger" sequencing after its inventor, Frederick Sanger) was amongst the first sequencing technologies to be developed. Due to ease of use and reliability, Sanger sequencing soon became the gold standard of sequencing technologies. However, there are limitations of Sanger-based sequencing which are problematic for metagenomic sequencing projects. The chain-terminator sequencing method is biologically biased [26], in that foreign DNA needs to be cloned into a bacterial vector (*Escherichia coli*). Thus, the DNA needs to be compatible with *E. coli* replication machinery. Sanger sequencing is also an expensive and low-throughput technology. As a result, Sanger-based metagenomic projects are often limited to sequencing fosmid and bacterial artificial chromosome libraries or low-diversity microbial communities.

Next-generation sequencing have overcome several of the disadvantages of Sanger sequencing. These include (1) cheaper cost per base sequencing, (2) substantially higher throughput, (3) simpler library preparation, and (4) no cloning step. However, working with next-generation sequencing data is not without its own new challenges. Each new technology has a different error model and biases that need to be considered during experimental design and sequence analysis.

3.2 Roche 454 Sequencing (GS20, GS FLX, GS FLX Titanium, GS FLX+)

Released in 2005, the Roche 454 platform implements the sequencing-by-synthesis approach. DNA templates are affixed to microbeads and amplified using emulsion PCR. Beads are then distributed into individual picoliter-sized pyrosequencing reactors. Nucleotide triphosphates are flowed across the plate in a specific sequence, with base incorporation marked by the release of pyrophosphate. The reads produced by this platform are significantly longer than that of Illumina, with the latest pyrosequencer from Roche (GS FLX+) producing reads up to 800 bp long. Error rates are also much lower than observed in Illumina sequences (0.49–1 %) [27, 28]. Many of these errors appear to be template context specific, with homopolymer tracts inducing insertion, deletion, and substitution errors in the reads. The "quality" scores are unlike those generated for Sanger in that they evaluate the probability that the homopolymer length at that base is correct [27]. Quality scores decrease regardless of whether the homopolymer is correct or not [27]. This makes setting quality thresholds difficult without any prior knowledge about the genome sequence. These medium-sized reads are appropriate for a range of applications, including resequencing [29], de novo assembly [30], transcriptomics [31], and metagenomics [32]. The yield from a 454 run is orders of magnitude lower than that of Illumina and thus is less applicable for applications that depend upon ultra-deep coverage. However, it has become a highly popular tool for sequencing of conserved marker gene tags, now commonly referred to as "pyrotag" sequencing [33]. The use of sample-specific labels facilitates the multiplexing of hundreds of samples in the one run.

3.3 Illumina Sequencing (Genome Analyzer I and II, HiSeq, and MiSeq)

Illumina has also adopted the sequencing-by-synthesis approach. The Illumina sequencing protocol begins by ligating template DNA to an adaptor sequence and thence onto a glass flow cell. The template DNA is subjected to bridge amplification, whereby each template is increased to roughly 1,000 copies. By using an isothermal polymerase and 3′ inactivated fluorescent nucleotides, Illumina is able to incorporate a solitary base each cycle. Each base addition is followed by an imaging step, which reads the fluorescent label. Single-base incorporation provides a huge advantage, as context-specific errors, such as those caused by homopolymers, are avoided and repetitive and low-complexity regions are easily sequenced over. The mean error rate per base of generated sequence is approximately between 1 and 2 %. This rate is at least ten times greater than that of Sanger sequencing. Error rates across the read are well characterized, with low rates at the 5′ end, progressing to much higher error rates in the 3′ end. While the occurrence of errors appears to be independent, the positive correlation between base position and error rate can give the impression of nonindependence in adjacent bases. Despite claims that sequence context-specific errors are avoided, 31–35 % of errors occur in the sequence after a guanine is incorporated [34]. Insertion-/deletion-based errors are extremely rare, with an error rate of less than 0.01 %. There are substantial biases against G+C- or A+T-rich sequences, most likely introduced by the DNA template amplification procedures [34–36]. In a metagenomic project, this bias can severely reduce the number of reads produced from species with nonoptimal G+C content. Fortunately, there appears to be some success in reducing the biases introduced by sequence preparation [36–38]. While primarily intended for resequencing projects, several studies have demonstrated the utility of Illumina short reads for de novo assembly, either by itself [39, 40] or in conjunction with other sequencing technologies [41, 42]. Features such as genome size, G+C%, and repetitive content will ultimately determine the feasibility of sequencing a genome using Illumina technology [43]. As the reads produced by Illumina become longer than the typical perfect-match repeat length in prokaryotic genomes [44], moderately complex metagenomes can yield complete or near-complete genomes using solely the Illumina platform [11].

3.4 Ion Torrent Personal Genome Machine (PGM)

While still a sequencing-by-synthesis technology, the Ion Torrent Personal Genome Machine (PGM) adopts a novel approach to DNA sequencing. The PGM overcomes the dependence on expensive photoreceptive equipment and artificial reagents by both using natural dNTPs and measuring the subtle changes in pH, as opposed to emitted light, during polymerization events [45]. The PGM is a compromise between the low throughput of 454 and ultrahigh-throughput Illumina but delivers the cheapest "minimum" run cost of the three platforms. As with the Roche 454 pyrosequencing approach, the PGM flows dNTPs over the reaction chamber, with zero or more of each nucleotide binding during each flow. During the polymerization reaction, a hydrogen ion is released, resulting in a decrease in the pH of the solution. The decrease in pH is proportionate to the number of nucleotides that bound during the flow. The PGM is estimated to have an error rate between 1 and 1.7 % [45–48], predominantly comprised of indels associated with over-/under-called flows [46–48]. Error rates increase by flow cycle and are nonhomogeneous across flows within the cycle [46]. The platform also introduces high-frequency indels (HFIs), where, at a given base in the reference, the majority of aligned reads share the same indel, suggesting a genuine difference between the sample and reference genomes [46, 47]. However, these "polymorphism-like" errors do not perpetuate across replicates of the same sample [46]. HFIs occur around every 1–2 kb relative to the reference sequence. As with Illumina, the PGM exerts coverage biases against DNA fragments with very low (≤20 %) or very high (≥80 %) G+C% content [46, 47], and this can result in misrepresentation of community members within the metagenome. Nevertheless, the PGM has been applied to amplicon libraries [49, 50] and metagenomic samples [51, 52].

With so many competing technologies available, platform distributors are constantly striving for longer, less error-prone, and greater numbers of reads. Sanger has had an undisputed reign as the sequencing gold standard for over two decades; however, the sheer volume of reads produced using next-generation platforms has seen these newer technologies supplant the old, especially for applications sensitive to low sequencing depth. Given the dominance of next-generation sequencing for metagenomic projects, the remainder of this review will focus on the analysis of metagenomes sequenced using Illumina and Roche 454 technologies.

4 Analysis of Metagenomic Data

Given the numerous variables that determine the nature of a metagenome, there is no universally applicable analysis strategy for all datasets. There are almost endless possibilities in terms of what information can be mined from metagenomic datasets. Assuming no restriction on the genes/species of interest, Fig. 1 illustrates a generic strategy for assembly, annotation, and analysis of a metagenome. In general, there are two workflows for analysis, and the choice of workflow is largely dependent on the sequencing saturation of the community. Since genomic read coverage is inversely proportional to the relative abundance of the species, it follows that in highly diverse, even communities, the metagenome assembly step is often bypassed in favor of "gene-centric" analysis approaches (left-hand side of Fig. 1). Clustering of reads to identify homologs can be performed either on the entire dataset prior to annotation, as illustrated in Fig. 1, or only on the subset of reads which do not have an annotation in public repositories.

Using Fig. 1 as a guide, the processes and a list of potential bioinformatics tools for performing each of these steps are discussed. While this is targeted at researchers who wish to implement a custom metagenome workflow, there are tools that provide whole metagenome annotation and comparative analysis [53–55].

5 Quality Assurance/Quality Control

Quality assurance methods are generally platform specific. While assembly methods often rely upon the dominant signal in the reads (i.e., correctly called bases), when there is low read coverage of the genome, such errors may prevent assembly or be perpetuated into the assembled contigs. To maximize chances of assembly, trimming and/or filtering of reads is a necessity. There are few published programs for filtering/trimming next-generation data (Table 1). This may be due to the lack of consensus and a degree of subjectivity, in the choice of filters/trims used on the data. Fortunately, the typical filters/trims used are not difficult to implement in scripting languages.

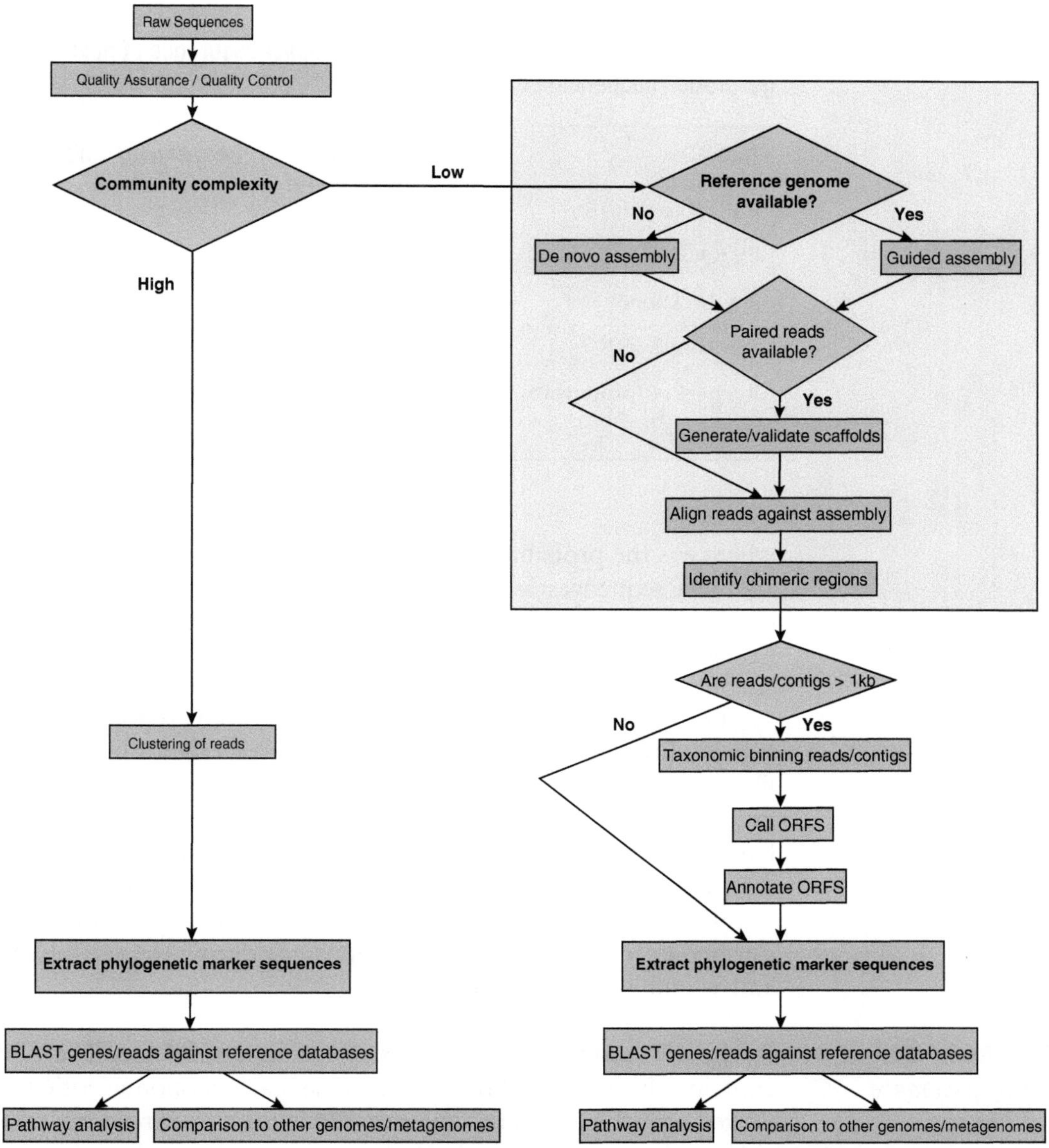

Fig. 1 Flow diagram describing the overall processes involved in analyzing metagenomic datasets. The workflow on the *left* corresponds to a "gene-centric" analysis, and the *right*, an "assembly based" analysis

5.1 Illumina Quality Assurance

Illumina reads are typically in FASTQ format, which combines both the nucleotide sequence and quality values into a single file. Each base in the read is assigned a quality score, and while the calculation of the quality score has changed through several iterations of CASAVA (Illumina sequencing software) development, the current quality scores (CASAVA 1.8) are identical to those of Sanger "Phred" scores. Each quality score is calculated as

$$-10 \times \log_{10} e$$

Table 1
Table of bioinformatics tools available for quality assurance of next-generation sequencing data

Tool	Sequencing platforms
FASTX toolkit [56]	All
FastQC [57]	Illumina
ngs_backbone [58]	All
Pyrobayes [59]	454
Roche 454 (proprietary)	454
Shore [60]	Illumina

where e is the probability that the given base is correct. Typically, Illumina sequences are trimmed on the 3′ side, but there is no definitive trim length. Inspection of the sequences with FastQC (Table 1) may help decide on appropriate trim choices. Vector and adaptor sequences should also be masked or trimmed from reads.

For mate-pair libraries, it has been commonly reported that a subset of the library will be paired end (often called the "shadow library"). This artifact is attributed to the mate-pair library preparation process. The shadow library will consist of read pairs which have the opposite orientation to the mate-paired reads, and the distance between pair members will be substantially smaller (~150 bp). Without a reference genome, these are difficult to filter out pre-assembly but will become apparent during the scaffolding and assembly validation steps.

5.2 Roche 454 Quality Assurance

Roche 454 sequencing output is stored in an SFF file. The SFF contains the flow readings observed as each mononucleotide is flowed over the template sequence. While these flow readings provide resolution to two decimal places, the actual number of incorporated bases is obtained by rounding these values to the nearest positive integer. Recent studies have shown that with GS FLX Titanium data, over-call/under-call error (often referred to as homopolymer error) increases both with distance from the read origin and with the length of the homopolymer in the template sequence [61]. Flow values which are intermediate between zero and one (in particular, 0.5–0.7) are also highly correlated with error [62]. Observing such a value early on in the reads may indicate loss of synchronicity in the templates on the bead, and as such these reads should be filtered from the dataset. Alternatively, truncation of reads prior to these flow values may also be considered. As with Illumina, linker and adaptor sequences should be filtered/trimmed from the reads.

Unlike Sanger or Illumina quality scores, Roche 454 quality scores are a measure of the confidence that the homopolymer length at the current base is correct [27]. It has been found that Roche 454 quality scores do *not* necessarily correlate well with low-quality regions. The incidence of ambiguous nucleotides (N's) and substitutions have been found to increase substantially towards the 3′ end of the read in GS FLX Titanium data [28]. From previous work, "N's" can occur during the same flow for numerous reads [63]. Consequently, we find it prudent to truncate reads before the occurrence of an "N."

6 Metagenomic Assembly

Sequence assembly from metagenomic data is highly sensitive to the read coverage for community members, and read coverage is largely determined by the community structure and sequencing volume. While the strategy outlined in Fig. 1 suggests that samples with high community diversity are unlikely to benefit from assembly, this entirely depends on the read depth achieved for the community. In addition, the types of algorithms used for different sequencing platforms (especially the short sequences of Illumina versus Sanger-based assemblers) may find the simultaneous de novo assembly of the entire community prohibitive, due to either computational expense or suboptimal assembly of individual species. With next-generation technologies, it is suggested that a separate assembly is conducted for each species of interest, with the parameters optimized for that species. With regard to guided assemblies, to our knowledge these are always performed in a species-by-species manner.

While we have considered the next-generation technologies in isolation, the complementary nature of these platforms can be leveraged through hybrid assembly approaches [41, 42]. Such an approach will often use the long-read library (e.g., Roche 454, Sanger) to generate initial contiguous sequences and then perform scaffolding using the high-volume Illumina paired-read libraries. The unique biases and error profiles for each platform allow the correction of 454/Sanger errors by taking a consensus of the Illumina reads.

6.1 Guided Assembly

Guided assemblies are preferable when sequence coverage is low (due to either low cellular abundance or high intraspecies variability) and especially when a reference genome from a closely related species (the same genus, say) is available. Table 2 shows a number of tools available for both guided and de novo assemblies.

While it may make sense to subtract the reads which are homologous to the reference genome from the metagenomic dataset, it is important to consider that some reads may represent

Table 2
Assemblers for next-generation sequence data. Hybrid assemblies denote assemblers which combine power of short (Illumina, SOLiD) and long reads (454, Sanger)

Assembler	Type	Sequencing platforms	Hybrid assembly	Library types	Metagenomic data?
ABBA [64]	Guided	Illumina	No	Unpaired	No
ABySS [65]	De novo	Illumina, SOLiD	No	Paired	No
ALLPATHS 2 ALLPATHS-LG [66]	De novo	Illumina	Yes	Paired, fragment	No
AMOS-cmp short-reads [67]	Guided	Illumina, 454	Yes	Paired, fragment	No
Euler-SR [68]	De novo	Illumina, 454	Yes	Paired, fragment	No
Meta-IDBA [69]	De novo	Illumina	No	Paired, fragment	Yes
MetaORFA [70]	De novo	All	Yes	Fragment	No
MetaVelvet [71]	De novo	Illumina, SOLiD	No	Paired, fragment	Yes
Minimus2 [72]	De novo	Any	Yes	Fragment	No
Mira3 [73]	De novo	Sanger, Illumina, 454, Ion Torrent, Pac Bio	Yes	Paired, fragment	No
Newbler (Roche)	De novo	454	No	Paired, fragment	No
Ray Meta [74]	De novo	Ion Torrent	No	Paired, fragment	Yes
Velvet [40]	De novo	Illumina, SOLiD	No	Paired, fragment	No

highly conserved regions across phylogenetically distant species and removing these could compromise assembly of other community members.

6.2 De Novo Assembly

De novo assembly of next-generation data requires a significantly different algorithmic approach to that used for the longer, and less erroneous, Sanger reads. Numerous short-read assemblers have been published (see Table 2); however, very few assemblers are designed to handle metagenomic data. The majority of assembly tools assume that the reads are derived from a single species. Assumptions about the genomic coverage distribution often restrict the algorithm to assembly of one species at a time. This is because low-coverage reads in these samples are deemed "errors" and high-coverage reads as "repetitive regions" and, as such, often ignored during the initial (and sometimes throughout) assembly. In the case of metagenomic data, these coverage differences could be due to a number of reasons, including dominant versus rare species, shared genes across species, G + C biases, or plasmid sequences. The basic process of assembly is that overlaps are identified between reads, and these overlapping read pairs are merged to form a contig (a "contiguous" sequence). Algorithms designed for Sanger reads typically use an overlap-layout-consensus approach, where the minimum "unit" is the read itself. Alignments were performed between read pairs, and sufficiently similar reads would be merged into a contig. By contrast, the majority of short-read assemblers initially deconstruct each read into a series of oligomers of a set "word" length (k, commonly referred to as "k-mers"). The word length is often a user-specified parameter, with longer word lengths generating more unique words, thus overcoming repetitive/nonunique regions in the metagenome. The compromise with longer word lengths is that the effective coverage is minimized, as overlapping reads do not necessarily start at the same position, and they may contain errors. Thus, the opposite is true of short k-mers, where the effective coverage is higher, but repetitive regions may limit contig extension and scaffolding. There are a large number of tools for assembling short-read data generated from the Illumina platform (Table 2). With regard to the Roche 454 platform, the long read lengths and the prevalence of insertion/deletion errors make it ill-suited for k-mer-based short-read assemblers. The Roche 454 proprietary assembler "Newbler" is generally recommended to assemble 454 data and has shown to resolve strain-level differences in metagenomes [75].

When assembling species from a metagenomic dataset, there is a substantial risk of forming chimeric assemblies. The risk of chimeras can be partially mitigated by using paired-read libraries, as one member of the pair may map to a common microbial sequence, such as a housekeeping gene or repetitive element, but the second member maps to a region outside the shared sequence. In complex microbial communities where there is a risk of chimeric assembly, using high identity thresholds (identity >95 %) should also reduce the chance of chimeras forming, as it is expected that sequence divergence and species-specific codon usage will result in sufficient variation in different species. Contigs should always be inspected for coverage breakpoints and abrupt changes in codon usage, G+C content, and other genomic signatures. Note that genome assembly is often an iterative procedure and may be improved by reassembling contigs (or the reads that they are comprised of) after binning.

7 Scaffolding

Scaffolding is the process whereby contigs are ordered and oriented relative to one another using paired-end or mate-paired sequences. While many assembly tools incorporate scaffolding into the assembly algorithm, it is always worth investigating whether further scaffolding can be achieved. In addition, scaffolding processes will indicate whether contigs are ambiguous within the assembly and also identify regions of the assembly which are potentially inconsistent. Given that very few assembly algorithms can handle multiple or hybrid read-pair libraries, post-assembly scaffolding tools may provide the only means to combine these datasets. The "shadow library" in Illumina mate-pair datasets will become apparent during scaffolding and may even confound this process. Fortunately, in the majority of cases, the "shadow library" does not dominate mate-pair libraries.

Excluding Bambus2, the majority of scaffolding tools are not explicitly "metagenome-aware" [76–78]. Bambus2 [79] achieves the "metagenome-scaffolding" specialization through an algorithm that attempts to detect biologically meaningful variants, simultaneously minimizing chimerism and improving scaffolding sensitivity.

8 Read Alignment for SNP Detection and Assembly Validation

In addition to aligning paired libraries against the assembly to generate scaffolds, there are other insights that can be gained by aligning reads (fragment or paired libraries) against the assembly. These insights include highlighting regions of low coverage, identifying local regions of misassembly, detecting repeated regions within the genome, and, finally, uncovering sequence-based variability between the individual genomes which contributed to the assembly. In the latter case, the detection of polymorphic sites via sequence alignment can then be used to delineate strains within the community [80].

Next-generation sequencing has revitalized the development of sequence alignment algorithms, with numerous algorithms to choose from, particularly for the short reads generated by Illumina [81, 82]. Aligners that handle Roche 454 data are less common, as they must tolerate indel errors within the reads [81].

In metagenomic assemblies, the mixed nature of the community can result in chimeric assemblies. There are no specific tools for detecting chimeras within a metagenomic assembly. However, by using a combination of assembly features, such as read-pair mappings, G+C skew changes, codon usage changes, or blasting ORFs found within contigs, chimeric regions can be detected.

9 Extracting Phylogenetic Marker Sequences

Phylogenetic marker genes are chosen such that they are ubiquitously present across species (often an indicator of "essential" function), yet have enough variability to provide species-specific sequences. Genes such as the small and large ribosomal subunit (SSU and LSU), *recA*, *rpoB*, *gyrB*, and *fusA* are commonly used phylogenetic markers. For strain-level differences or particular subgroups of prokaryotes, alternative markers may be more appropriate (e.g., *mcra* for methanogens). Contigs that contain a marker gene can be confidently assigned to a given species.

The SSU has been the most widely used phylogenetic marker for the past two decades. As such, there are several large sequence repositories available for annotating and comparing these genes [83–85]. While reads/contigs containing phylogenetic marker genes can be found with BLAST, the use of hidden Markov models (HMMs), which probabilistically describe the marker gene, tends to be more sensitive especially for partial gene sequences [86]. This is particularly relevant for the SSU gene, as its high copy number in prokaryotic genomes often causes breakpoints in sequence assembly. Contigs which cannot be associated with a phylogenetic marker sequence may be classified using sequence-"binning" methods.

10 Binning Metagenomic Sequences

Typically, binning aims to assign sequences (reads/contigs/scaffolds) to "bins" corresponding to their taxonomic origins. High-resolution binning (i.e., to the family/genus level) can be achieved with read or assembly fragments as small as 800 bp [87]. Smaller fragments may only be resolved to the phylum level, because they have too little signal to be classified with any accuracy [87–89]. Sequence-binning methods currently fall into broad categories, supervised and unsupervised binning (see Table 3 for a list of

Table 3
Software for "binning" reads/contigs

Tool	Sequence-binning approach	Supervised?
Carma3 [92]	Similarity	Yes
CompostBin [93]	Composition	No
MEGAN4 [94]	Similarity	Yes
PhyloPythiaS [95]	Composition	Yes
PhymmBL [87]	Composition	Yes
SGOM [96]	Composition	Semi-supervised
TACOA [97]	Composition	Yes
TaxSOM [98]	Composition	Semi-supervised
Taxy [99]	Composition	Yes
Tetra [88]	Composition	No
TreePhyla [100]	Composition	Yes

tools). Supervised methods either depend on direct comparisons between contigs and reference databases to classify sequences ("sequence similarity") or use species compositional features derived from reference databases to classify sequences ("sequence composition"). The limitation of these methods is that, at the present time, there is a poor representation of phylogenetic diversity across bacteria and archaea in public repositories. As a result, the majority of sequences will be poorly classified or spuriously assigned to unrelated taxa. However, the sequence-composition tools can often be retrained on biologically relevant data, such as large contigs generated from the sample under analysis.

In contrast, unsupervised binning is the process of using read features, such as G + C content, coverage, codon usage, oligonucleotide frequencies, and periodic sequence signatures [90], often in combination, to classify or cluster sequences into species "bins." Coarse features such as G + C content and read coverage can potentially separate very simple communities where members are distant or one species clearly dominates [9]. However, additional features may be required to accurately bin data from natural microbial communities. This can be achieved by using combinations of read features as input for computational and/or statistical classification algorithms. In addition, multiple coverage values can be derived for each contig by leveraging spatial, temporal, or even library construction differences across samples [91].

11 ORF (Open Reading Frame) Calling

When processing sequence data from a single species, ORF calling is generally attempted post-assembly. In addition to the previously mentioned assembly validation methods, identification of truncated or erroneous gene sequences through ORF calling can identify both adjacent contigs which have not been scaffolded and regions of potential misassembly, where genes appear truncated or spurious.

As is often the case in metagenomic assemblies, many of the genes within the assembly are only partial sequences. This requires ORF callers that can detect genes without penalizing reads that only contain partial gene sequences. Many of the tools for this purpose use machine-learning approaches, including HMMs and neural networks, to detect ORFs. Tools designed for ORF identification in metagenomic data have superior sensitivity to standard tools designed for complete gene sequences. GeneMark.hmm [101], MetaGene [102], FragGeneScan [103], and Orphelia [104] are all tools which are specifically designed for metagenomic data.

An alternative to using the machine-learning approaches list above is to query sequence databases for homologous genes or protein domains; however, the success of this approach relies upon the database containing relevant sequences. Sequence homology-based annotation approaches are discussed further below.

12 Read/ORF Annotation and Pathway Analysis

There are numerous sources of annotations for microbial sequences. These include the KEGG database [105], BioCyc [106], NCBI genomes, SEED [107], Pfam domain calling [108], and COG functional groupings [109]. Despite such abundant annotation resources, metagenomic datasets typically contain many novel sequences (>50 %), which have not been annotated [3]. In these instances, functional inference may be gleaned by identifying common sets of environmental variables under which novel sequences are found. The environmental sequence databases (such as CAMERA [53] and NCBI's environmental sequences) have numerous metagenomic datasets, with collection metadata, which may facilitate such analyses. The power of such analyses will grow as the size and diversity of samples available increase, but this will be contingent upon the quality of metadata recorded.

After gene annotation, evaluating the biochemical pathways present in the data can give a comprehensive understanding of community function. As is often the case with metagenomic data, the genes sampled from the community may only comprise partial biochemical pathways. This begs the question as to whether a partial pathway still conveys a biochemical function and potentially whether the absence is due to lack of sequencing depth or actually not found in the community. The use of tools such as MinPath [110] can be used to identify the minimum set of viable pathways present with a metagenomic community.

13 Comparison of Metagenomes

With the number and scale of metagenomic datasets increasing rapidly, the development of efficient analysis methods for multiple metagenomic datasets is essential. Initial methods provided only pair-wise comparison of samples and focused on detecting phylogenetic differences between samples [111, 112]. While this is broadly informative about differences in community composition, restricting the comparison to phylogeny gives little indication of functional differences between communities, especially with regard to novel species. More recently, tools have been developed which incorporate both functional and phylogenetic features and scale to multiple-sample comparison [55, 113, 114] (Table 4). However, very few of the currently available tools investigate the correlation between sample metadata and metagenome functional and phylogenetic features [115, 116].

Table 4
Tools for comparison and contrast of metagenomic sequences, often in combination with sample metadata

Tool	Number comparable	Use sample metadata
LEfSe [113]	Any	Yes
MEGAN4 [90]	Any	No
MG-RAST [55]	Any	No
ShotgunFunctionalizeR [111]	Any	No
STAMP [114]	Two	No
XIPE-TOTEC [112]	Any	No

14 Conclusion

Metagenomics provides a direct means to interrogate environmental sequences. While improvements in sequencing technology and culture-independent methods have increased the ease of generating these datasets, the bottleneck is, and will continue to be, the downstream analysis of such data.

Every metagenome is unique, warranting a customized approach to the community analyzed and the biological questions driving the study. Individual tool choice will depend on those best suited to the data, computing infrastructure and research objectives. Few tools are explicitly designed for metagenomic data, so understanding the assumptions made by algorithms is crucial to analysis outcomes.

In many instances, no single tool has been identified as unequivocally best for a given analysis procedure, and while tools classified as fulfilling the same purpose have overlapping function, they often can produce complimentary results. Thus, we encourage readers to compare and contrast the output of "equivalent" tools on their own datasets.

References

1. Amann R, Ludwig W, Schleifer K (1995) Phylogenetic identification and in situ detection of individual microbial cells without cultivation. Microbiol Rev 59:143–169
2. Breitbart M et al (2002) Genomic analysis of uncultured marine viral communities. Proc Natl Acad Sci U S A 99:14250–14255
3. Venter JC et al (2004) Environmental genome shotgun sequencing of the Sargasso Sea. Science 304:66–74
4. Breitbart M et al (2003) Metagenomic analyses of an uncultured viral community from human feces. J Bacteriol 185: 6220–6223
5. Hallam SJ et al (2004) Reverse methanogenesis: testing the hypothesis with environmental genomics. Science 305:1457–1462
6. Gill SR et al (2006) Metagenomic analysis of the human distal gut microbiome. Science 312:1355–1359
7. Warnecke F et al (2007) Metagenomic and functional analysis of hindgut microbiota of a wood-feeding higher termite. Nature 450: 560–565

8. Tringe SG et al (2005) Comparative metagenomics of microbial communities. Science 308:554–557
9. Tyson GW et al (2004) Community structure and metabolism through reconstruction of microbial genomes from the environment. Nature 428:37–43
10. Béjà O et al (2000) Bacterial rhodopsin: evidence for a new type of phototrophy in the sea. Science 289:1902–1906
11. Hess M et al (2011) Metagenomic discovery of biomass-degrading genes and genomes from cow rumen. Science 331:463–467
12. Hemme CL et al (2010) Metagenomic insights into evolution of a heavy metal-contaminated groundwater microbial community. ISME J 4:660–672
13. Pagani I et al (2012) The Genomes OnLine Database (GOLD) v. 4: status of genomic and metagenomic projects and their associated metadata. Nucleic Acids Res 40:D571–D579
14. Peterson J et al (2009) The NIH human microbiome project. Genome Res 19:2317–2323
15. Kroeber M et al (2009) Phylogenetic characterization of a biogas plant microbial community integrating clone library 16S-rDNA sequences and metagenome sequence data obtained by 454-pyrosequencing. J Biotechnol 142:38–49
16. Boetius A et al (2000) A marine microbial consortium apparently mediating anaerobic oxidation of methane. Nature 407:623–626
17. DeAngelis KM et al (2011) Characterization of trapped lignin-degrading microbes in tropical forest soil. PLoS ONE 6:e19306
18. Ding H, Valentine DL (2008) Methanotrophic bacteria occupy benthic microbial mats in shallow marine hydrocarbon seeps, Coal Oil Point, California. J Geophys Res 113:G01015
19. Edwards R et al (2006) Using pyrosequencing to shed light on deep mine microbial ecology. BMC Genomics 7:57
20. Havelsrud O et al (2011) A metagenomic study of methanotrophic microorganisms in coal oil Point seep sediments. BMC Microbiol 11:221
21. Poinar HN et al (2006) Metagenomics to paleogenomics: large-scale sequencing of mammoth DNA. Science 311:392–394
22. Turnbaugh PJ et al (2006) An obesity-associated gut microbiome with increased capacity for energy harvest. Nature 444:1027–1131
23. Coetzee B et al (2010) Deep sequencing analysis of viruses infecting grapevines: virome of a vineyard. Virology 400:157–163
24. Lazarevic V et al (2009) Metagenomic study of the oral microbiota by Illumina high-throughput sequencing. J Microbiol Meth 79:266–271
25. Qin J et al (2010) A human gut microbial gene catalogue established by metagenomic sequencing. Nature 464:59–65
26. Sorek R et al (2007) Genome-wide experimental determination of barriers to horizontal gene transfer. Science 318:1449–1452
27. Huse SM et al (2007) Accuracy and quality of massively parallel DNA pyrosequencing. Genome Biol 8:R143
28. Gilles A et al (2011) Accuracy and quality assessment of 454 GS-FLX Titanium pyrosequencing. BMC Genomics 12:245
29. Bordoni R et al (2008) Evaluation of human gene variant detection in amplicon pools by the GS-FLX parallel pyrosequencer. BMC Genomics 9:464
30. Moore M et al (2006) Rapid and accurate pyrosequencing of angiosperm plastid genomes. BMC Plant Biol 6:17
31. Hornshøj H et al (2009) Transcriptomic and proteomic profiling of two porcine tissues using high-throughput technologies. BMC Genomics 10:30
32. Jimnez DJ et al (2012) Structural and functional insights from the metagenome of an acidic hot spring microbial planktonic community in the Columbian Andes. PLoS ONE 7(12):e50269
33. Kunin V et al (2010) Wrinkles in the rare biosphere: pyrosequencing errors can lead to artificial inflation of diversity estimates. Environ Microbiol 12:118–123
34. Dohm JC et al (2008) Substantial biases in ultra-short read data sets from high-throughput DNA sequencing. Nucleic Acids Res 36:e105
35. Hillier LW et al (2008) Whole-genome sequencing and variant discovery in C. elegans. Nat Meth 5:183–188
36. Aird D et al (2011) Analyzing and minimizing PCR amplification bias in Illumina sequencing libraries. Genome Biol 12:R18
37. Quail MA et al (2008) A large genome center's improvements to the Illumina sequencing system. Nat Meth 5:1005–1010
38. Kozarewa I et al (2009) Amplification-free Illumina sequencing-library preparation facilitates improved mapping and assembly of (G+C)-biased genomes. Nat Meth 6:291–295
39. Dohm JC et al (2007) SHARCGS, a fast and highly accurate short-read assembly algorithm for de novo genomic sequencing. Genome Res 17:1697–1706
40. Zerbino DR, Birney E (2008) Velvet: algorithms for de novo short read assembly using de Bruijn graphs. Genome Res 18:821–829
41. DiGuistini S et al (2009) De novo genome sequence assembly of a filamentous fungus using Sanger, 454 and Illumina sequence data. Genome Biol 10:R94
42. Reinhardt JA et al (2009) De novo assembly using low-coverage short read sequence data from the rice pathogen Pseudomonas syringae pv. oryzae. Genome Res 19:294–305

43. Whiteford N et al (2005) An analysis of the feasibility of short read sequencing. Nucleic Acids Res 33:e171
44. Kassai-Jáger E et al (2008) Distribution and evolution of short tandem repeats in closely related bacterial genomes. Gene 410:18–25
45. Rothberg JM et al (2011) An integrated semiconductor device enabling non-optical genome sequencing. Nature 475:348–352
46. Bragg LM et al (2013) Shining a light on dark sequencing: characterising errors in ion torrent PGM data. PLoS Comp Biol 9(4):e1003031
47. Quail MA et al (2012) A tale of three next generation sequencing platforms: comparison of ion torrent, pacific biosciences and illumina MiSeq sequencers. BMC Genom 13:341
48. Loman NJ et al (2012) Performance comparison of benchtop high-throughput sequencing platforms. Nat Biotech 30(5):434–439
49. Jünemann S et al (2013) Bacterial community shift in treated periodontitis patients revealed by ion torrent 16S rRNA gene amplicon sequencing. PLoS ONE 7(8):e41606
50. Yergeau E et al (2012) Next-generation sequencing of microbial communities in the Athabasca river and its tributaries in relation to oil sands mining activities. Appl Environ Microbiol 78(21):7626–7637
51. Solonenko SA et al (2013) Sequencing platform and library preparation choices impact viral metagenomes. BMC Genomics 14:320
52. Whitely AS et al (2012) Microbial 16S rRNA Ion Tag and community metagenome sequencing using the Ion Torrent (PGM) Platform. J Microbiol Meth 91:80–88
53. Seshadri R et al (2007) CAMERA: a community resource for metagenomics. PLoS Biol 5:e75
54. Markowitz VM et al (2006) An experimental metagenome data management and analysis system. Bioinformatics 22:e359–e367
55. Meyer F et al (2008) The metagenomics RAST server - a public resource for the automatic phylogenetic and functional analysis of metagenomes. BMC Bioinformatics 9:386
56. The Hannon Lab FASTX toolkit. http://hannonlab.cshl.edu/fastx_toolkit/index.html
57. Babraham Bioinformatics FASTQC. *FASTQC* at http://www.bioinformatics.bbsrc.ac.uk/projects/fastqc/
58. Blanca J et al (2011) ngs_backbone: a pipeline for read cleaning, mapping and SNP calling using Next Generation Sequence. BMC Genomics 12:285
59. Quinlan AR et al (2008) Pyrobayes: an improved base caller for SNP discovery in pyrosequences. Nat Meth 5:179–181
60. Ossowski S et al (2008) Sequencing of natural strains of Arabidopsis thaliana with short reads. Genome Res 18:2024–2033
61. Balzer S et al (2010) Characteristics of 454 pyrosequencing data-enabling realistic simulation with flowsim. Bioinformatics 26:i420–i425
62. Quince C et al (2009) Accurate determination of microbial diversity from 454 pyrosequencing data. Nat Methods 6:639–641
63. Bragg LM et al (2012) Fast, accurate error-correction of amplicon pyrosequences using Acacia. Nat Methods 9(5):425–426
64. Salzberg SL et al (2008) Gene-boosted assembly of a novel bacterial genome from very short reads. PLoS Comput Biol 4:e1000186
65. Simpson JT et al (2009) ABySS: a parallel assembler for short read sequence data. Genome Res 19:1117–1123
66. MacCallum I et al (2009) ALLPATHS 2: small genomes assembled accurately and with high continuity from short paired reads. Genome Biol 10:R103
67. Chaisson MJ, Pevzner PA (2008) Short read fragment assembly of bacterial genomes. Genome Res 18:324–330
68. Pop M et al (2004) Comparative genome assembly. Brief Bioinform 5(3):237–248
69. Peng Y et al (2011) Meta-IDBA: a De Novo assembler for metagenomic data. Bioinformatics 27(13):i94–i101
70. Ye Y, Tang H (2009) An ORFome assembly approach to metagenomics sequences analysis. J Bioinform Comput Biol 7: 455–471
71. Namiki T et al (2012) Metavelvet: an extension of Velvet Assembler to *de novo* metagenome assembly from short sequence reads. Nucleic Acids Res 40(20):e155
72. Treangen TJ et al (2011) Next generation sequence assembly with AMOS. Curr Protoc Bioinform 33:11.8.1–11.8.18
73. Chevreux B, Wetter T, Suhai S (1999) Genome sequence assembly using trace signals and additional sequence information. Computer Sci Biol 99:45–56
74. Boisvert S et al (2012) Ray Meta: scalable de novo metagenome assembly and profiling. Genome Biol 13:R122
75. Morowitz MJ et al (2011) Strain-resolved community genomic analysis of gut microbial colonization in a premature infant. Proc Natl Acad Sci U S A 108:1128–1133
76. Bonfield JK, Whitwham A (2010) Gap5—editing the billion fragment sequence assembly. Bioinformatics 26:1699–1703
77. Boetzer M et al (2011) Scaffolding pre-assembled contigs using SSPACE. Bioinformatics 27:578–579
78. Salmela L et al (2011) Fast scaffolding with small independent mixed integer programs. Bioinformatics 27:3259–3265
79. Koren S, Treangen TJ, Pop M (2011) Bambus 2: scaffolding metagenomes. Bioinformatics 27:2964–2971
80. Eppley J et al (2007) Strainer: software for analysis of population variation in community genomic datasets. BMC Bioinformatics 8: 398

81. Li H, Durbin R (2010) Fast and accurate long-read alignment with Burrows–Wheeler transform. Bioinformatics 26:589–595
82. Langmead B et al (2009) Ultrafast and memory-efficient alignment of short DNA sequences to the human genome. Genome Biol 10:R25
83. Cole JR et al (2009) The ribosomal database project: improved alignments and new tools for rRNA analysis. Nucleic Acids Res 37:D141–D145
84. DeSantis TZ et al (2006) Greengenes, a chimera-checked 16S rRNA gene database and workbench compatible with ARB. Appl Environ Microbiol 72:5069–5072
85. Pruesse E et al (2007) SILVA: a comprehensive online resource for quality checked and aligned ribosomal RNA sequence data compatible with ARB. Nucleic Acids Res 35:7188–7196
86. Huang Y, Gilna P, Li W (2009) Identification of ribosomal RNA genes in metagenomic fragments. Bioinformatics 25:1338–1340
87. Brady A, Salzberg SL (2009) Phymm and PhymmBL: metagenomic phylogenetic classification with interpolated Markov models. Nat Meth 6:673–676
88. Teeling H et al (2004) TETRA: a web-service and a stand-alone program for the analysis and comparison of tetranucleotide usage patterns in DNA sequences. BMC Bioinformatics 5:163
89. McHardy AC et al (2007) Accurate phylogenetic classification of variable-length DNA fragments. Nat Meth 4:63–72
90. Mrázek J (2009) Phylogenetic signals in DNA composition: limitations and prospects. Mol Biol Evol 26:1163–1169
91. Albertsen M et al (2013) Genome sequences of rare, uncultured bacteria obtained by differential coverage binning of multiple metagenomes. Nat Biotech 31:533–538
92. Gerlach W, Stoye J (2011) Taxonomic classification of metagenomic shotgun sequences with CARMA3. Nucleic Acids Res 39:e91
93. Huson DH et al (2011) Integrative analysis of environmental sequences using MEGAN4. Genome Res 21:1552–1560
94. Chatterji S et al (2008) CompostBin: a DNA composition-based algorithm for binning environmental shotgun reads. Res Comput Mol Biol Proc 4955:17–28
95. Patil KR et al (2011) Taxonomic metagenome sequence assignment with structured output models. Nat Meth 8:191–192
96. Chan C-K et al (2008) Binning sequences using very sparse labels within a metagenome. BMC Bioinformatics 9:215
97. Diaz N et al (2009) TACOA - Taxonomic classification of environmental genomic fragments using a kernelized nearest neighbor approach. BMC Bioinformatics 10:56
98. Weber M et al (2011) Practical application of self-organizing maps to interrelate biodiversity and functional data in NGS-based metagenomics. ISME J 5:918–928
99. Meinicke P, Aßhauer KP, Lingner T (2011) Mixture models for analysis of the taxonomic composition of metagenomes. Bioinformatics 27:1618–1624
100. Schreiber F et al (2010) Treephyler: fast taxonomic profiling of metagenomes. Bioinformatics 26:960–961
101. Besemer J, Borodovsky M (1999) Heuristic approach to deriving models for gene finding. Nucleic Acids Res 27:3911–3920
102. Noguchi H, Park J, Takagi T (2006) MetaGene: prokaryotic gene finding from environmental genome shotgun sequences. Nucleic Acids Res 34:5623–5630
103. Rho M, Tang H, Ye Y (2010) FragGeneScan: predicting genes in short and error-prone reads. Nucleic Acids Res 38:e191
104. Hoff K et al (2008) Gene prediction in metagenomic fragments: a large scale machine learning approach. BMC Bioinformatics 9:217
105. Kanehisa M, Goto S (2000) KEGG: Kyoto Encyclopedia of genes and genomes. Nucleic Acids Res 28:27–30
106. Karp PD et al (2005) Expansion of the BioCyc collection of pathway/genome databases to 160 genomes. Nucleic Acids Res 33:6083–6089
107. Overbeek R et al (2005) The subsystems approach to genome annotation and its use in the project to annotate 1000 genomes. Nucleic Acids Res 33:5691–5702
108. Finn RD et al (2010) The Pfam protein families database. Nucleic Acids Res 38:D211–D222
109. Tatusov R et al (2003) The COG database: an updated version includes eukaryotes. BMC Bioinformatics 4:41
110. Ye Y, Doak TG (2009) A parsimony approach to biological pathway reconstruction/inference for genomes and metagenomes. PLoS Comput Biol 5:e1000465
111. Huson DH et al (2007) MEGAN analysis of metagenomic data. Genome Res 17:377–386
112. Lozupone C, Knight R (2005) UniFrac: a new phylogenetic method for comparing microbial communities. Appl Environ Microbiol 71:8228–8235
113. Kristiansson E, Hugenholtz P, Dalevi D (2009) ShotgunFunctionalizeR: an R-package for functional comparison of metagenomes. Bioinformatics 25:2737–2738
114. Rodriguez-Brito B, Rohwer F, Edwards RA (2006) An application of statistics to comparative metagenomics. BMC Bioinformatics 7:162
115. Segata N et al (2011) Metagenomic biomarker discovery and explanation. Genome Biol 12:R60
116. Parks DH, Beiko RG (2010) Identifying biologically relevant differences between metagenomic communities. Bioinformatics 26:715–721

Chapter 16

Targeted Genomics of Flow Cytometrically Sorted Cultured and Uncultured Microbial Groups

Sophie Mazard, Martin Ostrowski, Ross Holland, Mikhail V. Zubkov, and David J. Scanlan

Abstract

High throughput sequencing of genetic material recovered from environmental samples (i.e., metagenomics) is becoming the method of choice for either medical or environmental genomic studies. However, the large amount of data and complexity of the sequenced "biomes" present challenges for teasing meaningful results out of the mass. Here, we describe a targeted genomic pipeline which uses fluorescence-activated cell sorting (FACS) in combination with multiple displacement amplification (MDA) of nucleic acids that allows to dissect a complex system into its component parts to facilitate high-quality single-cell, or targeted population, genomic reconstructions of microbial communities. This pipeline is presented with methods for collecting, concentrating, and preserving cells from aquatic and marine environments suitable for flow cytometric processing at a later date.

Key words Flow-cytometry, Genomics, Whole-genome amplification, Marine *Synechococcus*, Cyanobacteria, Single cell genomics, Metagenomics, Population genomics

1 Introduction

The study of environmental microbiology has been revolutionized by the application of culture-independent techniques and dramatic improvements in sequencing technologies that have led to a wave of new discoveries [1–3]. High-throughput sequencing of DNA extracted directly from environmental samples, known as metagenomics, offers the ability to reconstruct the metabolic potential of a community as a whole, and thereby infer metabolic function [4] without the need to cultivate organisms. While this approach has been very successful for systems displaying relatively low microbial diversity, such as the microbial communities functioning in acid mine drainage biofilms or solar salterns [4, 5], the ability to sequence more complex microbial communities to near completion is inconceivable. An early example of the power and limitations of metagenomics applied to the marine microbial community [1] demonstrated that

Ian T. Paulsen and Andrew J. Holmes (eds.), *Environmental Microbiology: Methods and Protocols*, Methods in Molecular Biology, vol. 1096, DOI 10.1007/978-1-62703-712-9_16, © Springer Science+Business Media, LLC 2014

large scaffolds of genomic DNA could be reconstructed for the two most abundant members, but the majority of sequences obtained could not be assembled nor assigned to any specific microorganism (*see* also [6]). Complexity is the norm for the vast majority of microbial communities found in soil, marine, or gut environments where larger and/or less abundant cells have the potential to make significant contributions to particular functional components of the ecosystem, for example, CO_2 fixation [7–9] and N_2 fixation [10], or DMSP degradation [11]. Since the smallest most abundant cells dominate the DNA of metagenomic libraries, whole community sequencing runs the risk of presenting an incomplete or biased overview of the system.

In the context of systems biology the functioning of an ecosystem or biogeochemical cycle is the sum of its parts. Additional tools are therefore required to achieve an overall view of the processes and players that contribute to the system as a whole. A number of promising approaches have emerged to address this issue, often by applying reductionist approaches to dissect out distinct groups, single cells or to focus in on groups of organisms involved in specific activities. For example, DNA stable isotope probing [12–14], is a method that enables the fractionation of community DNA which has incorporated a stable isotope from a labeled substrate to establish links between metabolic function and microbial identify. The emergence of single cell genomics may also help to uncover the metabolic functions of many organisms that are relatively rare in the environment but are influential in important biogeochemical processes [15].

Here we describe a methodology to dissect out specific groups of photosynthetic marine organisms from complex communities using flow sorting. Whole-genome amplification (WGA) of DNA by Multiple Displacement Amplification (MDA) is used to recover sufficient DNA from sorted cells for subsequent high-throughput sequencing. Unicellular photosynthetic organisms can be discriminated from heterotrophic organisms as well as each other on the basis of fluorescence from their photosynthetic pigments (i.e., chlorophyll and phycoerythrin) and cell size using flow cytometry [16, 17]. However, this approach is not restricted to organisms with distinct cytometric signatures but can be adapted to other sample types or target specific community members through the use of fluorescent in-situ hybridization [18], fluorescently tagged antibodies [19], fluorescence-based activity assays or by monitoring population changes during controlled environmental perturbations [20, 21].

2 Materials

Prepare all solutions using analytical grade reagents and ultrapure water (pre-filtered through 0.22 μm pore-size filters and deionized to attain a sensitivity of 18.2 MΩ cm at 25 °C). Ideally, equipment for environmental sampling should be sterilized, but this is not always practical when working in the field. As an alternative, equipment can be washed with hot water (>60 °C) and neutral detergent (e.g., 1.0 % v/v neutracon, Decon Laboratories Ltd, Hove, UK), then rinsed thoroughly with hot water to remove traces of detergent before a final rinse with Milli-Q water (Millipore Ltd, Watford, UK).

2.1 Environmental Sample Concentrating

1. Sampling device (e.g., acid-washed polycarbonate sample bottles).
2. Peristaltic pump with multiple heads (e.g., Watson Marlow 323S/D Pump Drive with one 313D pump head and capacity for additional 313X extension pump heads).
3. Peristaltic pump tubing, platinum-cured silicone, e.g., Masterflex L/S 18 3.2 mm internal diameter, 6.4 mm outer diameter.
4. Large bore platinum-cured silicone tubing, e.g., Masterflex L/S 36 for swinnex pre-filter assembly.
5. Luer lock hose barb adapters and suitable barbed reducing connectors (e.g., Cole-Parmer SN-30800-02, SN-06458-50).
6. CellTraps™, CT22 or CT40, Mem-Teq Ventures, UK, www.mem-teq.co.uk
7. Syringes and cryotubes.
8. Liquid nitrogen, dry shipper or liquid nitrogen dewar.

Optional

9. Nylon mesh pre-filter to exclude large particles and debris: 20–60 μm pore size.
10. Swinnex filter holders, 47 mm diameter.
11. Nucleopore track etched polycarbonate filter membrane 10.0 μm pore size, 47 mm diameter.

2.2 Fluorescence-Activated Cell Sorting

1. Flow sorting of phytoplankton from seawater concentrates was carried out with a MoFlo flow cytometer (DAKO-Cytomation, UK) equipped with a blue laser (488 nm, Coherent Sapphire™, Coherent), and a tunable ultraviolet argon-ion laser (405 nm, Innova 90c, Coherent, UK).
2. Calibration beads: 3.0 μm diameter yellow-green fluorescent microspheres (Fluoresbrite®, cat. 17155-2, Polysciences, Germany). Store at 4 °C and prepare fresh working stock at regular intervals.
3. SYBR Green I Nucleic Acid stain (Invitrogen, Carlsbad, CA, USA), 300 mM sodium citrate. Store in the dark at −20 °C. Prepare fresh staining solutions and pre-filter (0.2 μm pore size) before use.

4. Sterile sample tubes.
5. Disposable filters (0.2 μm pore size, e.g., Minisart, Sartorius Stedim UK Ltd, Surrey, UK) and syringes.
6. Bleach solution: 5.0 % v/v sodium hypochlorite.
7. Ethanol, 70 % v/v.
8. Neutracon®, 1 % v/v (Decon® Laboratories Ltd, Hove, UK).
9. Concentrated sheath fluid: sodium chloride, 10 % w/v solution, sterile. The sheath fluid for sorting (0.1–1.0 % w/v NaCl) is prepared by diluting the sterile 10.0 % w/v NaCl stock into pre-filtered (0.2 μm diameter pore size) Milli-Q water.
10. Ice/dry ice/liquid nitrogen.

2.3 Whole-Genome Amplification and Quantitation

1. Clean sterile environment to prepare the amplification reaction, e.g., UV laminar flow.
2. Powder free gloves.
3. Set of pipettors to dispense volumes from 1 to 100 μL, with sterile DNA/RNA free filter tips.
4. Thermal cycler and sterile PCR tubes.
5. Illustra GenomiPhi v2 DNA amplification kit (GE Healthcare, Little Chalfont, UK). Store at −80 °C.
6. Quant-IT™ PicoGreen® dsDNA assay kit (Invitrogen, Carlsbad, CA, USA). Store at 4 °C.
7. Fluorescence 96-well microplate (e.g., Greiner Bio-One Fluotrac 96-well black microplate, VWR, UK).
8. Nuclease-free sterile water (e.g., Nuclease-free water, Ambion®, UK).
9. Fluorescence microplate reader with the following filters: excitation ~480 nm (i.e., fluorescein), emission ~520 nm (ex. 485 nm/em. 535 nm is suitable).

3 Methods

3.1 Collection, Concentration, and Storage of Environmental Samples for Later Processing

Environmental sampling for flow cytometry typically involves prefiltration and size fractionation (*see* Fig. 1) in order to target the microbial community within a specific size range and to avoid blocking the cytometer sort nozzle with large cells, aggregates and large particles. Preservation of cells in the aqueous phase, rather than on solid supports is preferable and offers better recovery after cryopreservation.

1. Seawater samples collected from Niskin bottles, or other suitable samplers, are immediately filtered by gravity filtration through a nylon mesh pre-filter (20 μm or 100 μm pore size) to

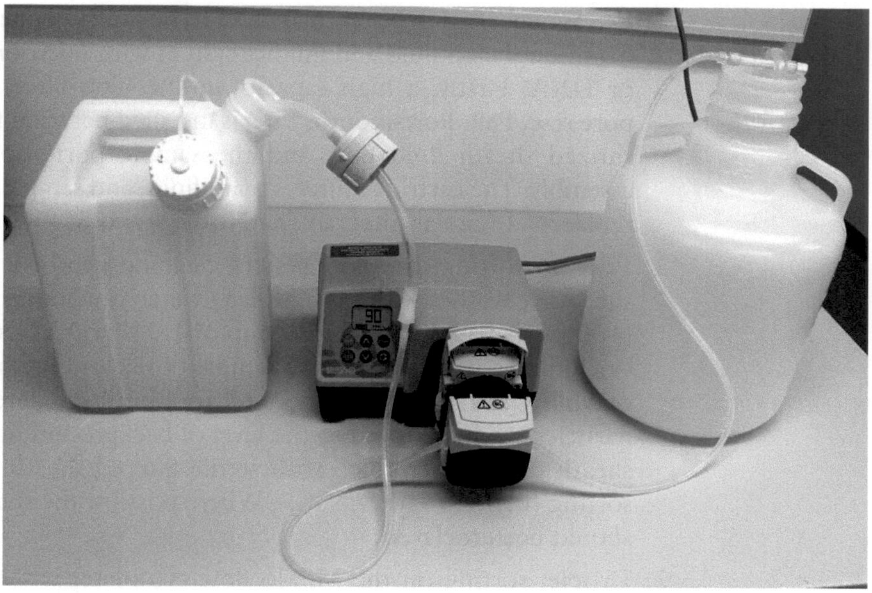

Fig. 1 Pre-filtration and size fractionation setup for the preparation of environmental cells

remove large particles, such as microplankton or mesoplankton, respectively, and collected in acid-washed polycarbonate bottles (*see* **Note 1**). Concentration of the microbial biomass is carried out by pumping the sample through a CellTrap™ (CT22 or CT40) 0.22 μm hollow fiber filter at a moderate pump speed (70–90 rpm) to avoid excessive back-pressure and preserve the integrity and fluorescence of the collected cells (*see* **Note 2**, that could be checked by flow cytometry). With this setup it has been possible to concentrate up to 3 L of seawater within 20 min, depending on the amount of biomass in the original sample.

2. When sufficient sample has been concentrated the biomass is extracted from the hollow fiber CellTrap™ according to the manufacturer's instructions, transferred to a labeled cryotube and flash frozen in liquid nitrogen without the addition of preservative (*see* **Note 3**).
3. Individual CellTraps™ can be used for additional rounds of concentration and biomass extraction of the same environmental sample although the filtration efficiency decreases.
4. Flash-frozen samples can be transferred to −80 °C freezer for long-term storage (beyond 2 years).

3.2 Cell Sorting by Flow Cytometry

The key to successful enrichment by flow cytometry is to eliminate or minimize sources of cellular and DNA contamination from the sheath fluid and within the system.

1. Precautions need to be taken in order to minimize contamination from the sheath fluid and fluidics system with microorganisms or DNA. Firstly, a high capacity in-line ultra-filter (0.1 μm pore size, Pall, Portsmouth, UK) is installed between the pressurized sheath fluid tank and the sample port—sort block assembly. The cartridge filter, sheath fluid, and sample lines are replaced at regular intervals (*see* **Notes 4** and **5**).
2. Prior to sorting, the flow system cleaning protocol involves successive washes with fresh Milli-Q water at working pressure (~60 psi, 1 h) followed by 20 min of 1.0 % (v/v) Neutracon® (including a back flush of the sample port), and further rinses of Milli-Q water and 20 % v/v ethanol until the detection of particles can be reduced to the level of background noise. The sheath tank is then filled with sterile filtered sheath fluid for sorting (0.1–1.0 % w/v NaCl). Where possible the sheath tank should be autoclaved.
3. Droplet sorting on the MoFlo flow cytometer is carried out with the highest purity settings ("sort-purify" and "single-drop" sort mode) with a 70 μm nozzle, a nozzle amplitude of ~12.5 V, and a drop-drive frequency of ~95,000 Hz. The drop delay is calibrated manually before each sorting session by using 3.0 μm diameter yellow-green fluorescent microspheres sorted onto a glass slide followed by counting the beads using epi-fluorescence microscopy.
4. The preserved samples can be partially thawed on ice and an aliquot (100–200 μL) of the sample removed with a sterile spatula while the sample was still frozen (*see* **Note 6**). Depending on the number of cells in the concentrate it is likely that the sample will need to be diluted 5–10× in pre-filtered sterile artificial seawater or suitable buffer (*see* [22] for a range of suitable seawater recipes) and the sample flow rate will need be adjusted in order to achieve a balance between the number of particles detected and the sort-rate of the target population.
5. Phycoerythrin-containing *Synechococcus* can be differentiated from other cells by their fluorescent pigment signatures using bivariate dot plots (Summit Software version 3.1, Dako-Cytomation, UK) of forward scatter and orange fluorescence (580 ± 30 nm band-pass filter). They can be further resolved from other groups, such as *Prochlorococcus* and photosynthetic pico-eukaryotes using bivariate plots of red fluorescence (670 ± 30 nm band-pass filter) with forward scatter and orange fluorescence. Two successive sort purifications are required to purify *Synechococcus*, particularly from environmental samples where *Prochlorococcus* and small heterotrophic bacteria outnumber the target cells by more than tenfold. The first sort enriches for *Synechococcus* (~300–500k cells) and reduces the

number of detectable *Prochlorococcus* 1,000-fold, as determined by flow cytometry and by PCR amplification of the 16S-23S rRNA internal transcribed spacer region (ITS) which varies in length between *Synechococcus* and *Prochlorococcus* (*see* **Note 7**).

6. Prior to the second sort, perform a rapid DNA-staining with SYBR Green I (10 min) in the presence of 30 mM (final concentration) sodium citrate [23]. This second step enables the discrimination of non-pigmented, small heterotrophic bacteria and *Prochlorococcus* cells displaying dim red fluorescence, which would otherwise not be detected, based on DNA fluorescence (530 ± 15 nm band-pass filter).
7. Twice-sorted target cells (10–50k cells at ~1,000 cells/μL) can be flash frozen in liquid N_2 or on dry ice and stored at −80 °C until further processing.

3.3 Whole-Genome Amplification and Quantitation

1. In order to minimize additional processing steps that could result in the loss of material or the introduction of contaminating nucleic acids the sorted cell suspensions are used directly as template for whole-genome amplification (*see* **Note 8**). At this stage care must be taken and the procedure must be carried out in "nucleic acid free" environment. The use of a UV laminar flow hood, sterilized equipment, and powder free gloves is recommended.
2. The amplification reaction is carried out with Phi 29 polymerase using the Illustra GenomiPhi v2 DNA amplification kit according to the manufacturer's instructions (*see* **Note 9**). Briefly, 10 μL of sample (~10,000 cells, *see* **Note 10**) is used per reaction, mixed with 9.0 μL of sample buffer supplied in the kit (to a total volume of 19.0 μL) and subjected to a short thermal denaturation step of 3 min at 95 °C (*see* **Note 11**). The sample is cooled rapidly to 4 °C before addition of the reaction buffer and enzyme mix (1.0 μL of enzyme plus 9.0 μL of sample buffer) while kept at 4 °C. The amplification reaction is carried out for 2 h 15 min at 30 °C and the terminated by a 10 min incubation step at 65 °C. With this approach the highest yield of DNA was obtained from samples that received the minimum thermal denaturation (3 min) at 95 °C (*see* **Note 12**).
3. Quantification of the amplification products is carried out in triplicate on 10- to 100-fold dilutions of the reaction (target concentration around 1–10 ng/μL) against a standard curve prepared from a DNA sample of known concentration using the dsDNA-specific fluorescent dye PicoGreen (*see* **Note 13**). Briefly, using black fluorescence 96-well microplates, a standard curve is produced from serial dilutions of a lambda DNA standard (provided in the PicoGreen assay kit) alongside the samples to be quantified (e.g., 1 μL of the diluted sample diluted in 99 μL of TE buffer) (*see* **Note 14**). The freshly prepared dye

solution (100 μL of 1× concentration, *see* **Note 15**) is mixed with each sample and incubated for 3–5 min at room temperature before analysis with a fluorescence microplate reader using with excitation ~480 nm and emission ~520 nm.

4. The obtained amplified product can then be used directly for high throughput sequencing or other applications (*see* **Note 16**). For some applications, e.g., fosmid library construction, it might be necessary to carry out further enzymatic treatments (*see* [24]) or a purification step to remove leftover oligonucleotide primers and Phi29 enzyme from the amplification reaction.

4 Notes

1. Additional size fractionation can be carried out with a polycarbonate nucleopore filter (e.g., 3.0, 5.0, or 10.0 μm pore size).
2. For best results, air bubbles should be excluded from entering the CellTrap™.
3. Alternate chemical free fixations methods have been also used, for example a method using microwave treatment (*see* [25]).
4. During periods of inactivity the fluidics lines and sheath container were thoroughly flushed with fresh Milli-Q water (at least 1 h) followed by 20 % ethanol (at least 1 h), and stored with 20 % ethanol to prevent the growth of microorganisms.
5. Decontamination of other types of flow cytometer can be carried out using a similar process although the decontamination protocol should be adapted for the hardware in use and be rigorously tested using culture and molecular techniques. For example see the BD application note "Decontamination of the BD FACSAria II System using the: Prepare for aseptic sort procedure" 2008 (www.bdbiosciences.com/documents/FacsariaII_Decontamination.pdf).
6. In this way it is possible to avoid defrosting the entire sample and use only the required amount necessary to sort the desired number of target cells.
7. The Internal Transcribed Spacer (ITS) region was amplified using the primer set 16S-1247F and 23S-241R as described in [26].
8. The volume of nucleic acid template used can be up to 10 μL, though it is recommended to use as small a volume as possible. Larger volumes impact dramatically the yield of nucleic acid retrieved in the following amplification.
9. To minimize random amplification spike produced by the Phi29 polymerase, it is recommended to carry out multiple replicate amplification reactions which can then be pooled before further use.

10. The manufacturer recommends a minimum template input of 10 ng DNA in 1.0 μL volume into the amplification reaction. A maximum yield of ~1,000 cells/μL can be attained by flow sorting at the highest drop-drive frequencies which produce the smallest droplets. Based on an average genome size of 2.5 Mb the theoretical yield of DNA contained within 10,000 *Synechococcus* cells is ~25 fg. This yield is significantly below the lowest recommended limit. However, an average of 2–3 μg of amplified dsDNA was recovered with such input.
11. For sensitive material, the initial heat denaturation step within the whole-genome amplification can be replaced by a chemical denaturation step as described in the kit protocol.
12. The initial heat denaturation step must not be extended beyond the recommended 3 min. Increased heat denaturation results in a dramatic reduction in the yield of the final amplified product, due to breakage of the nucleic acid strands into shorter fragments or nicking of the strands that are not optimal for the Phi29 polymerase.
13. Determination with the usual absorbance at 260 nm, e.g., using a NanoDrop cannot be used due to the presence of saturating quantities of nucleotides and random priming hexamers in the reaction buffers. These interfere with the reading producing inaccurate concentration estimations.
14. Ensure that the solutions used to prepare the dilutions and blanks are prepared with nuclease and nucleic acid free water.
15. The PicoGreen fluorescent dye is light sensitive, so ensure that it is protected from light by wrapping tubes with foil or using light excluding microplate covers. The fluorescent dye is conserved in DMSO, which becomes solid at 4 °C. The stock solution of dye should be thawed at room temperature for about 20 min before use to ensure the whole volume of dye solution is dissolved before diluting in TE buffer. Do not overheat the solution to accelerate the process as it may damage the dye stock. Once developed the fluorescence signal remains stable for a few hours at room temperature.
16. It is recommended that quality controls are carried out to verify that no contamination from exogenous bacteria and/or nucleic acid have been introduced during flow sorting. These can be performed by general 16S PCR amplification followed by Restriction Fragment Length Polymorphism, Denaturing Gradient Gel Electrophoresis [27] or any other suitable molecular analyses at each stage of the process. Quality control should also be performed to ensure that no loss or bias was introduced in the targeted community during cell sorting through flow cytometry (e.g., creating a mixture of different targeted cultured isolates and analyzing every stage of the processing as an environmental sample using a specific marker, *see* [28]).

Acknowledgements

This work was supported by grants from the FP6 EU Marine Genomics Network and NERC grants NE/F004249/1 to DJS and MVZ and NE/D003385/1 to DJS.

References

1. Venter JC et al (2004) Environmental genome shotgun sequencing of the Sargasso Sea. Science 304:66–74
2. Shendure J, Ji H (2008) Next-generation DNA sequencing. Nat Biotechnol 26:1135–1145
3. Rothberg JM, Leamon JH (2008) The development and impact of 454 sequencing. Nat Biotechnol 26:1117–1124
4. Tyson GW et al (2004) Community structure and metabolism through reconstruction of microbial genomes from the environment. Nature 428:37–43
5. Cuadros-Orellana S et al (2007) Genomic plasticity in prokaryotes: the case of the square haloarchaeon. ISME J 1:235–245
6. Tringe SG et al (2005) Comparative metagenomics of microbial communities. Science 308:554–557
7. Li WKW (1994) Primary production of prochlorophytes, cyanobacteria, and eukaryotic ultraphytoplankton: measurements from flow cytometric sorting. Limnol Oceanogr 39:169–175
8. Jardillier L et al (2010) Significant CO_2 fixation by small prymnesiophytes in the subtropical and tropical northeast Atlantic Ocean. ISME J 4:1180–1192
9. Cuvelier ML (2010) Targeted metagenomics and ecology of globally important uncultured eukaryotic phytoplankton. Proc Natl Acad Sci U S A 107:14679–14684
10. Montoya JP et al (2004) High rates of N2 fixation by unicellular diazotrophs in the oligotrophic Pacific Ocean. Nature 430:1027–1032
11. Howard EC et al (2008) Abundant and diverse bacteria involved in DMSP degradation in marine surface waters. Environ Microbiol 10:2397–2410
12. Radajewski S et al (2000) Stable-isotope probing as a tool in microbial ecology. Nature 403:646–649
13. Dumont MG, Murrell JC (2005) Stable isotope probing - linking microbial identity to function. Nat Rev Microbiol 3:499–504
14. Chen Y, Murrell JC (2010) When metagenomics meets stable-isotope probing: progress and perspectives. Trends Microbiol 18:157–163
15. Stepanauskas R, Sieracki ME (2007) Matching phylogeny and metabolism in the uncultured marine bacteria, one cell at a time. Proc Natl Acad Sci U S A 104:9052–9057
16. Olson RJ, Zettler ER, DuRand MD (1993) Phytoplankton analysis using flow cytometry. In: Kemp PF, Sherr BF, Sherr EB, Cole JJ (eds) Handbook of methods in aquatic microbial ecology. Lewis Publishers, Boca Raton, FL, pp 175–186
17. Palenik B et al (2009) Coastal *Synechococcus* metagenome reveals major roles for horizontal gene transfer and plasmids in population diversity. Environ Microbiol 11:349–359
18. Yilmaz S et al (2010) Fixation-free fluorescence in situ hybridization for targeted enrichment of microbial populations. ISME J 4:1352–1356
19. Rivers AR, Jakuba RW, Webb EA (2009) Iron stress genes in marine *Synechococcus* and the development of a flow cytometric iron stress assay. Environ Microbiol 11:382–396
20. Zubkov MV et al (2001) Linking the composition of bacterioplankton to rapid turnover of dissolved dimethylsulphoniopropionate in an algal bloom in the North Sea. Environ Microbiol 3:304–311
21. Shi Y, McCarren J, DeLong EF (2012) Transcriptional responses of surface water marine microbial assemblages to deep-sea water amendment. Environ Microbiol 14:191–206
22. Moore LR et al (2007) Culturing the marine cyanobacterium *Prochlorococcus*. Limnol Oceanogr Meth 5:353–362
23. Marie D et al (1997) Enumeration and cell cycle analysis of natural populations of marine picoplankton by flow cytometry using the nucleic acid stain SYBR Green I. Appl Environ Microbiol 63:186–193
24. Zhang K et al (2006) Sequencing genomes from single cells by polymerase cloning. Nat Biotechnol 24:680–686
25. Mary I et al (2010) Metaproteomic and metagenomic analyses of defined oceanic microbial populations using microwave cell fixation and flow cytometric sorting. FEMS Microbiol Ecol 74:10–18
26. Rocap G et al (2002) Resolution of Prochlorococcus and Synechococcus ecotypes by using 16S-23S ribosomal DNA internal transcribed spacer sequences. Appl Environ Microbiol 68:1180–1191
27. Schäfer H et al (2002) Genetic diversity of 'satellite' bacteria present in cultures of marine diatoms. FEMS Microbiol Ecol 42:25–35
28. Mazard S et al (2012) Multi-locus sequence analysis, taxonomic resolution and biogeography of marine *Synechococcus*. Environ Microbiol 14:372–386

Chapter 17

Quantitative Microbial Metatranscriptomics

Scott Gifford, Brandon Satinsky, and Mary Ann Moran

Abstract

The direct retrieval and sequencing of environmental RNA is emerging as a powerful technique to elucidate the in situ activities of microbial communities. Here we provide a metatranscriptomic protocol describing environmental sample collection, rRNA depletion, mRNA amplification, cDNA synthesis, and bioinformatic analysis. In addition, the preparation of internal RNA standards and their addition to the sample are discussed, providing a method by which transcript numbers can be expressed as absolute abundances in the environment and more readily compared to other biogeochemical and ecological measurements.

Key words Metatranscriptome, rRNA depletion, mRNA amplification, Sequencing, Subtractive hybridization

1 Introduction

Advances in molecular techniques have revolutionized the field of microbial ecology, particularly in revealing the extraordinary phylogenetic and functional diversity contained within microbial communities. A major contemporary challenge is identifying which components of this complex functional gene pool are actively being expressed and how that expression varies over time and space. The direct collection and sequencing of RNA from the environment (termed metatranscriptomics) fulfill this need by providing a measure of a community's instantaneous transcriptional response to its surrounding environment. The development of this method in parallel with advances in next-generation sequencing technologies has made metatranscriptomics a powerful approach for analyzing in situ microbial expression in a wide variety of habitats.

The metatranscriptomics approach was first described by Poretsky et al. [1], and while there have been several modifications since then, it consists largely of the same modules (Fig. 1). Cellular biomass is rapidly collected from the environment in a manner that

Ian T. Paulsen and Andrew J. Holmes (eds.), *Environmental Microbiology: Methods and Protocols*, Methods in Molecular Biology, vol. 1096, DOI 10.1007/978-1-62703-712-9_17, © Springer Science+Business Media, LLC 2014

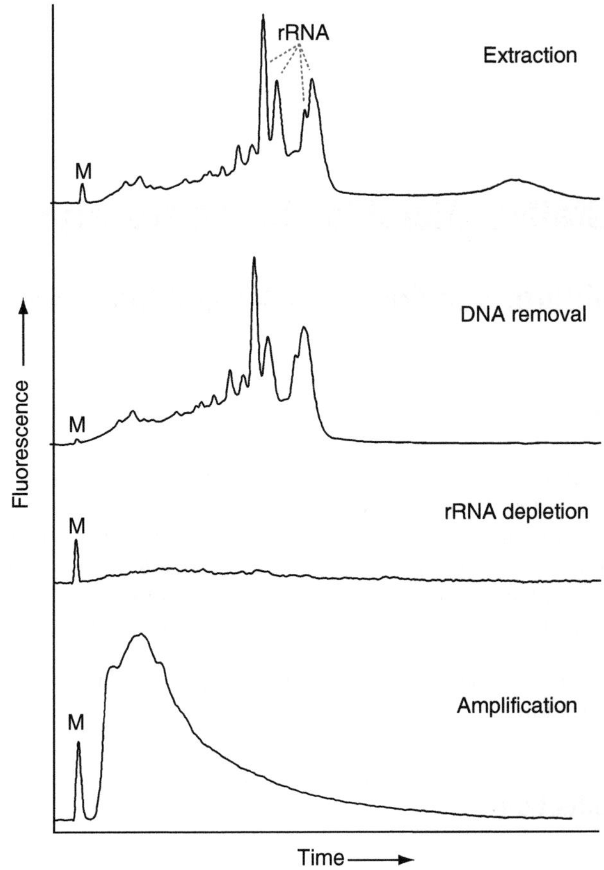

Fig. 1 Size distributions of sample RNA at different stages of processing visualized with the Experion automated gel electrophoresis system. Fragment length and abundance are proportional to run time and fluorescence, respectively. The gel marker is labeled "M." The distinct rRNA peaks dominate the total RNA pool in the extracted and TurboDNase-treated samples but are greatly diminished after subtractive hybridization

disturbs ambient conditions as little as possible. RNA is extracted from the samples and treated with DNase to remove any residual DNA. As ribosomal RNA (rRNA) makes up the majority of total cellular RNA, steps are taken to decrease rRNA abundance in order to increase the yield of messenger RNA (mRNA) sequences in the resulting libraries. Partially due to rRNA reduction, there is a significant decrease in total RNA mass and the sample can be linearly amplified to produce sufficient material for sequencing. Finally, the amplified RNA (aRNA) is converted to double-stranded cDNA, which can be sequenced through a variety of methods (Sanger, 454 pyrosequencing, Illumina, etc.). Using this approach, Poretsky et al. [1–3] and others [4–9] were able to successfully characterize metatranscriptomes from a variety of environments.

The most significant methodological challenge for metatranscriptomics has been efficient removal of rRNA. Poretsky et al. [1–3] used a dual-removal approach based on two commercially available kits. In the first round, rRNA is enzymatically digested by an exonuclease that targets the 5′ monophosphates found on rRNA, leaving mRNAs, which have a 5′ triphosphate, intact.

In the second round, biotinylated probes are hybridized to the rRNA and are bound to streptavidin-coated magnetic beads, allowing for physical separation via a magnetic stand. Typically, this dual approach removes ~50 % of contaminating rRNA [2, 9], although concern has been raised that the first round may cause bias in the resulting transcript library [10]. More recently, Stewart et al. [11] improved rRNA removal efficiency by using only the second (hybridization based) approach but creating custom hybridization probes to target the rRNAs in each individual sample. This method has been shown to decrease the proportion of rRNA reads to 10–30 % of total sequences ([11, 12], and Satinsky unpublished data).

A second challenge has been the interpretation of transcript abundances, which have traditionally been measured only as relative proportions within a sample. The ability to make quantitative interpretations, including cross sample comparisons, is limited when only proportional data is available. For example, a change in the abundance of one transcript category in a metatranscriptome causes the other categories' proportional representations to change, even if the absolute abundance of those other types remains constant [9, 13]. This limitation can be overcome by the addition of a standard (an artificial mRNA) just prior to starting the sample extraction [9, 13, 14]. Since both the amount of standard added and the amount of standard recovered are known, one can calculate the depth of sequencing and absolute copy number of a transcript category in a more ecologically relevant unit, such as copies/volume or copies/mass.

Here we present an updated version of the Poretsky et al. [1] protocol, using the custom subtractive hybridization protocol developed by Stewart et al. [11] for rRNA removal, and discuss the addition of internal standards to obtain absolute copy numbers in the environment [9, 14]. The method takes advantage of several commercial kits, and the reader should thoroughly familiarize himself/herself with each kit's manual. A number of steps use spin cartridges for purification, which efficiently capture mRNA-sized fragments (>200 nt), but small RNAs, including many regulatory RNAs, are likely not retained. Quantification is carried out spectrophotometrically (e.g., Nanodrop spectrophotometer) or with a fluorescence assay (e.g., PicoGreen for DNA and RiboGreen for RNA). Nucleic acid size distributions are visualized with an Agilent Bioanalyzer or Experion automated gel electrophoresis system. Care needs to be taken when dealing with RNA samples (*see* **Note 1**).

2 Materials

2.1 Materials for Environmental Collection

Peristaltic or vacuum pump.

Tubing (preferably acid washed).

Prefilter (if desired, for example a 3 μm pore size) and collection filter (typically 0.22 μm pore size) (*see* **Note 2**).

Filter housings.

Liquid nitrogen or RNALater (Applied Biosystems, Austin, TX).

Graduated 10 or 20 L carboy.

Sterile forceps.

Whirl-Pak® bags (Nasco, Fort Atkinson, WI).

2.2 Materials for RNA Extraction

Vortex station with 50 mL tube adapter (MO BIO Laboratories, Carlsbad, CA).

Rubber mallet and scissors.

50 and 15 mL Falcon tubes.

RNeasy RNA extraction kit (Qiagen, Valencia, CA).

Extra RLT buffer (Qiagen, Valencia, CA).

β-mercaptoethanol.

30 mL syringe and 18–21 gauge needles.

Centrifuge for both large (50 and 15 mL) and small (1.5 mL) tubes.

100 % molecular grade ethanol.

Vacuum manifold.

0.2 mm low-binding zirconium beads (OPS Diagnostics, Lebanon, NJ). Sterilized by heating at 500 °C overnight in a combustion oven.

2.3 Materials for Removal of Residual DNA

Turbo DNA-free (Applied Biosystems, Austin, TX).

Centrifuge.

Incubator.

2.4 Materials for Ribosomal RNA Reduction

T7 modified PCR primers (*see* Stewart et al. [11] for primer design).

Herculase II Fusion Polymerase (Agilent Technologies, Santa Clara, CA).

QIAquick PCR purification kit (Qiagen, Valencia, CA).

MEGAscript Transcription Kit (Applied Biosystems, Austin, TX).

MEGAclear Kit (Applied Biosystems, Austin, TX).

Biotin-11-CTP (10 mM) (Roche Applied Science, Indianapolis, IN).

Biotin-16-UTP (10 mM) (Roche Applied Science, Indianapolis, IN).

SUPERase•In RNase Inhibitor (Applied Biosystems, Austin, TX).

RNeasy MinElute Cleanup Kit (Qiagen, Valencia, CA).

Streptavidin-coated Magnetic Beads (New England Biolabs, Ipswich, MA).

20× Sodium Chloride-Citrate (SSC) Buffer (RNase-free) (Applied Biosystems, Austin, TX).

DynaMag Spin Magnet (Invitrogen, Carlsbad, CA).

Formamide (100 %).

0.1 M NaOH (nuclease free).

2.5 Materials for mRNA Amplification

MessageAmp™ II-Bacteria aRNA Amplification Kit (Applied Biosystems, Austin, TX).

Thermal cycler.

Incubator.

Tabletop centrifuge (all centrifugations are conducted at ~9,400 × *g*).

100 % ethanol.

2.6 Materials for cDNA Synthesis

Universal RiboClone cDNA Synthesis System (Promega, Madison, WI).

Incubator or hybridization oven.

Refrigerated incubator or thermal cycler.

0.1 mM nuclease-free EDTA (Applied Biosystems, Austin, TX).

QiaQuick PCR cleanup kit (Qiagen, Valencia, CA).

Vacuum manifold.

Centrifuge.

3 Methods

3.1 Environmental Collection

While a variety of methods can be used to collect biomass for RNA extraction, important points to consider are that the sample is collected as quickly as possible to prevent turnover of the mRNA pool and to keep the sampling conditions as close to ambient as possible to reduce transcriptional response to changes during collection. The sample should be preserved immediately after collection, either by snap freezing in liquid N_2 or by addition of an appropriate preservative. For optimal downstream processing, attempt to collect enough biomass to yield 5–20 μg of total RNA. Here we describe a collection method for aquatic environments.

1. Set up the filtration system consisting of tubing, prefilter (optional), 0.22 μm filter, and a graduated carboy (Fig. 2).
2. Place one end of the tubing in the water, and draw water through the filter system, measuring the volume filtered by its accumulation in the collection carboy. The appropriate volume to filter will depend on the environmental cell concentration. For coastal or limnological samples, 5–10 L is often sufficient.

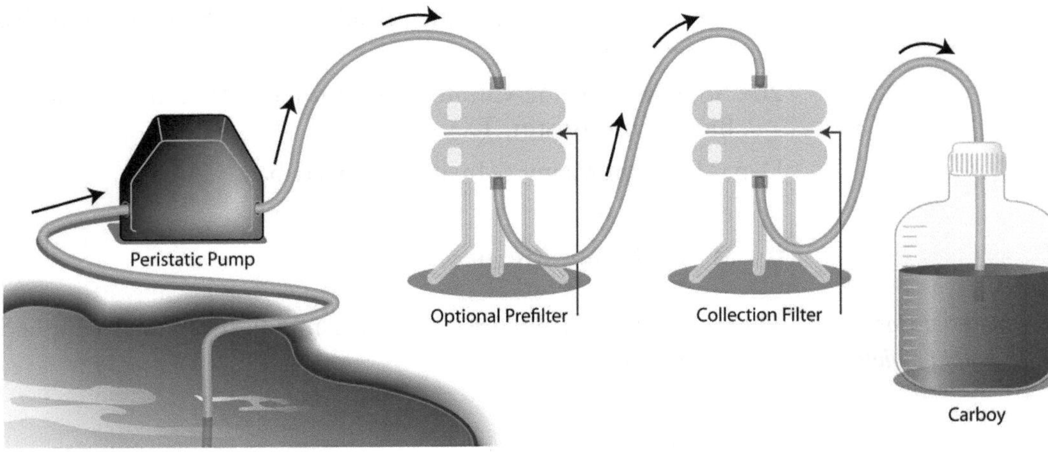

Fig. 2 Filtration setup for direct cell collection from an aquatic environment

Oligotrophic samples may require higher volumes. Total collection times should be kept as short as possible, optimally finishing the collection in 5–10 min and no longer than 30 min (*see* **Note 3**).

3. After the desired volume has been filtered, allow any water remaining in the line to pass through the filters. For optimal RNA yield, the surface of the filter should be nearly dry.
4. Fold the 0.22 μm filter and place into a Whirl-Pak® bag. Remove any air from the Whirl-Pak® by squeezing it out with your gloved fingers. Place the Whirl-Pak® into a liquid N_2 dewar. Alternatively, preserve the filter by submerging it in a tube containing 10 mL RNALater (Applied Biosystems, Austin, TX).
5. Repeat the process to collect an additional filter to be used for the DNA extraction; this is needed for the custom probe rRNA reduction protocol [11].

3.2 Internal Standard Synthesis

Construction of the internal RNA standards is done by in vitro transcription of DNA templates; these can be either commercially available plasmids (such as is commonly used for cloning) or synthesized DNA that is inserted into a plasmid. The use of a template that is already part of commercially available plasmids is attractive for its ease of use and low cost [9]. However, these plasmids make size customization difficult and often contain regions of homology to functional proteins. An alternative approach is to create a custom sequence, which is then synthesized and inserted into a plasmid, providing optimal control over sequence length and composition [14]. For either approach, each final plasmid construct should contain the following components (in order): an RNA polymerase promoter sequence, a unique internal standard sequence (*see* **Note 4**), and a restriction site (targeting a unique site in the vector and preferably producing a blunt end).

The appropriate amount of internal standard added to a sample is based on the expected total RNA mass yield. An addition of 0.5 % proportion of internal standard mass to expected sample mass of RNA is appropriate for next-generation sequencing. For a detailed protocol of standard construction using commercially available plasmids *see* Gifford et al. [9], or for custom-designed and synthesized standards *see* Satinsky et al. [14].

3.3 RNA Extraction

Many different methods are available for RNA extraction depending upon the environment of interest (aquatic, terrestrial, tissue, etc.). Here, we describe a modified approach based on Qiagen's RNeasy kit.

1. Prepare a 50 mL Falcon tube with 8 mL RLT buffer (β-mercaptoethanol added) and 2 mL beads.
2. Add the internal RNA standards to each Falcon tube. Each standard should be added independently (i.e., not as a pooled master mix) so that pipetting errors will be included in variance estimates.
3. Remove the filter from liquid nitrogen or −80 °C storage, break up to expose the most filter surface, and add to the Falcon tube (*see* **Note 5**). After adding the filter pieces to the Falcon tube, cap tightly and seal with parafilm. If the samples were preserved using RNALater the filters should be removed from the RNALater solution and any excess RNALater allowed to drip off the filter by gently squeezing with sterile forceps. The filter should then be placed into a Whirl-Pak bag, snap frozen in liquid nitrogen, and processed as described above.
4. Place the Falcon tubes on a vortex adapter and vortex at maximum speed for 10 min.
5. Centrifuge at 2,300 ×*g* for 1 min.
6. Using a 1,000 μL pipette, transfer the liquid to a clean 15 mL Falcon tube. Ideally, 80–90 % of the original volume should be recovered.
7. Centrifuge at 2,300 ×*g* for 5 min.
8. Gently pour the supernatant into a clean 50 mL Falcon tube, being careful not to disturb the pellet. At this point, the supernatant should be free of all beads and filter material.
9. Add 1× volume of 100 % ethanol.
10. Shear the sample by drawing the ethanol–lysis mixture up into the 30 mL syringe with an 18–21 gauge needle and then expelling. Repeat three times, then draw up the solution, and keep it in the 30 mL syringe.
11. Place an RNeasy spin cartridge on the vacuum manifold. Turn on the vacuum, and slowly expel the lysis mixture from the syringe into the cartridge (*see* **Note 6**). After all the lysate has been filtered, remove the column from the manifold, placing

it back in the collection tube, and centrifuge at 10,000 ×*g* for 1 min to remove any residual lysate solution.

12. Continue by following the standard RNeasy protocol as described in the kit manual. Briefly, wash once with 700 μL RW1 and twice with 500 μL RPE. Conduct a final centrifugation to remove any residual solutions. Place in a new collection tube, and elute with two separate aliquots of 50 μL RNase-free water. Place on ice.
13. Quantify the RNA yield with either a Nanodrop spectrophotometer or RiboGreen-based fluorometric technique.

This is a potential stopping point. The eluted RNA can be frozen at −80 °C. However, it is a good idea to keep the number of freeze/thaws to a minimum to reduce RNA degradation, so if possible continue on with the DNA removal step.

3.4 Removal of Residual DNA

A double treatment with TurboDNase is highly effective in digesting contaminant DNA in the RNA preparation. Note that for this and all other air incubations, place the tube(s) in a rack that allows ample air movement around the tube. For many of the reactions, it is important that temperature is uniform around the tube.

1. The sample should be in 90 μL of nuclease-free water.
2. Add 10 μL DNase buffer and 3 μL TurboDNase.
3. Incubate at 37 °C for 20 min in an incubator.
4. Remove the mixture from the incubator, and add an additional 3 μL of TurboDNase.
5. Return to the incubator for another 20 min.
6. Add 20 μL inactivation reagent, and incubate at room temperature for 2 min, vortexing every 20 or 30 s.
7. Centrifuge at max speed (typically 18,400 ×*g*) for 1 min.
8. Being careful not to disturb the inactivation reagent at the bottom of the tube, transfer the supernatant (~90–100 μL) to a new tube and place on ice.

This is a potential stopping point. Store the sample at −80 °C

3.5 Ribosomal RNA Reduction

Here we provide a brief overview of the custom rRNA depletion protocol and direct the reader to the original description by Stewart et al. [11] for specific details. This method uses broad-specificity primers to PCR amplify rRNA genes from a DNA sample collected in parallel with the RNA samples (the DNA filter must be extracted prior to starting the rRNA subtraction protocol). Several independent amplifications are carried out to generate probe sets for different contaminating rRNA molecules to be removed (i.e., 16S/23S bacteria, 16S/23S archaea, 18S/28S eukaryotes). The universal primers are modified to incorporate a T7 promoter into the PCR products. The PCR-amplified rDNA templates are then transcribed in vitro to make antisense rRNA probes containing biotinylated nucleotides. The probes are hybridized to the sample rRNA, bound to streptavidin magnetic beads, and physically separated from the rest of the sample via a magnetic stand.

3.5.1 PCR Amplification of rRNA Genes

1. For each rRNA gene amplification (16S bacterial, 23S bacterial, 16S archaeal, 23S archaeal, 18S eukaryotic, and 28S eukaryotic), prepare 4–5 individual 50 μL PCR reactions in 0.2 mL tubes on ice. For each reaction, combine 5–100 ng template DNA, 10 μL Herculase 5× buffer, 0.5 μL dNTP (100 mM), 1.25 μL forward primer (10 μM), 1.25 μL reverse primer (10 μM), 1 μL Herculase II Fusion Polymerase, and nuclease-free water to 50 μL reaction volume. Mix samples and briefly centrifuge.
2. Place the reactions in a thermal cycler, and run with one of the two following protocols. For all targets other than bacterial 23S, denature at 92 °C for 2 min; run 35–40 cycles at 95 °C for 20 s, 55 °C for 20 s, and 72 °C for 2 min; and end with a final extension at 72 °C for 3 min. For bacterial 23S targets, denature at 92 °C for 2 min; run 35–40 cycles of 95 °C for 20 s, 39 °C for 20 s, and 72 °C for 90 s; and end with a final extension at 72 °C for 3 min.
3. Pool the replicate 50 μL reactions into a single microcentrifuge tube.
4. Clean up the reactions using a QIAquick PCR purification kit, eluting in 30–50 μL of elution buffer (EB).
5. Quantify the PCR products with Nanodrop spectrophotometer or PicoGreen-based fluorometric method. It is critical to obtain 250–500 ng/μL of pooled PCR products before proceeding to the in vitro transcription (*see* **Note** 7).

3.5.2 Biotin-Labeled Antisense RNA Probe Creation

6. Antisense rRNA probes are synthesized via in vitro transcription with T7 RNA polymerase using the MEGAscript High Yield Transcription kit. Prepare separate 20 μL reactions for each rRNA probe type (16S, 18S, etc.). In vitro transcription reaction volumes can be doubled to increase yield if necessary.
7. For each rRNA gene product (16S, 18S, etc.), combine the following in order in a 0.2 mL PCR tube: 1 μL PCR amplicons (250–500 ng) from previous amplification, 2 μL ATP (75 mM), 2 μL GTP (75 mM), 1.5 μL CTP (75 mM), 1.5 μL UTP (75 mM), 3.75 μL biotin-11-CTP (10 mM), 3.75 μL biotin-16-UTP (10 mM), 2 μL 10× buffer, 0.5 μL RNase Inhibitor (Ambion), and 2 μL T7 polymerase.
8. Incubate in a thermal cycler at 37 °C overnight (heated lid set to 105 °C).

9. Add 1 μL DNase I to each reaction and incubate for 30 min at 37 °C in a thermal cycler.
10. Clean up the reaction with a MEGAclear kit, eluting in 50 μL of elution buffer.
11. Quantify probe concentration using either a Nanodrop spectrophotometer or a RiboGreen-based fluorescence method. A good transcription will result in >50–75-fold increase over the input DNA mass.

3.5.3 Subtractive Hybridization

12. Determine the input quantities of sample RNA template and biotinylated rRNA probes. Ideally, 250–500 ng of sample RNA (i.e., the original RNA pool containing mRNA and rRNA) will be used in the rRNA reduction processes. However, subtraction can be successfully conducted using lower template quantities if necessary. Each individual probe should be added at a probe:template ratio of 2:1. For example, if 500 ng of sample RNA is added to the reaction, then 1,000 ng of each unique probe should be added to the same reaction. The final total RNA (sample + rRNA probes) in the depletion reaction is calculated as (sample RNA mass) + [(number unique rRNA probes) × (2 × sample RNA mass)].
13. Calculate the volume of streptavidin bead suspension required and prepare by washing. A volume of 100 μL of streptavidin beads can be used with up to 2,000 ng total RNA (rRNA probes + RNA sample). Based on the total RNA calculated in **step 12**, add the appropriate volume of streptavidin beads needed into a 1.5 mL tube. Place the tube in a magnetic stand and let sit for 3 min. Discard the supernatant. Remove the tube from the stand, and resuspend the beads in an equal volume of 0.1 M NaOH. Place back on the stand, bind the beads, and discard the supernatant. Remove from the stand, and add an equal volume of 1× SSC buffer to the beads, mixing thoroughly to resuspend. Again separate the beads, and discard the supernatant. Repeat the 1× SSC wash twice, and on the third wash leave the beads in the SSC buffer and place on ice.
14. In a 0.5 mL tube combine RNA sample and each rRNA probe (volumes determined in **step 12**), 1 μL RNase inhibitor, 2.5 μL 20× SSC buffer, and 10 μL 100 % formamide. Bring the volume up to 50 μL with water.
15. Incubate in a thermal cycler for 5 min at 70 °C followed by ramping down to 25 °C using 5 °C increments for 1 min each.
16. Remove from the thermal cycler, and incubate for 5 min at room temperature. During this period it is useful to continue on with the bead dry-down step below (**step 17**).
17. Place the washed streptavidin beads (**step 13**) on the magnetic stand, and allow the beads to separate for 3 min. Discard the supernatant.

18. To the hybridization reaction tube add 1× SSC–20 % formamide solution so that the end volume of the hybridization reaction is equal to the initial aliquoted bead volume (**step 13**). For example, if the initial volume of beads aliquoted to deplete an individual reaction was equal to 200 μL, then add 150 μL of 1× SSC–20 % formamide solution to the 50 μL hybridization reaction.
19. Add the hybridization reaction mix from **step 18** to the tube containing the dried beads (**step 17**). Incubate at room temperature for 10 min, occasionally flicking to mix.
20. Place the tube in a magnetic stand, and allow the beads to separate for 3 min.
21. Transfer the supernatant (containing the purified RNA sample) into a clean 1.5 mL collection tube.
22. Resuspend the beads with 1× SSC, matching the original volume of the bead suspension (**step 13**). Return the beads to the stand and incubate for 3 min. Transfer the supernatant to the tube containing the first aliquot of supernatant (**step 21**).
23. Clean up and concentrate the depleted RNA using an RNeasy MinElute kit (Qiagen).
24. Quantify the enriched mRNA, and confirm rRNA reduction with a Bioanalyzer or an Experion system. This is a potential stopping point. Store at −80 °C.

3.6 mRNA Amplification

To obtain enough material for sequencing, the enriched mRNA sample is typically linearly amplified using the MessageAmp™ II-Bacteria aRNA Amplification Kit, consisting of four main steps: polyadenylation, reverse transcription to single-stranded cDNA, second-strand cDNA synthesis, and in vitro transcription to antisense aRNA. The user should closely read and follow the protocol described in the kit manual. Here we only provide a brief overview.

3.6.1 Polyadenylation

1. Add 10–200 ng of mRNA in a total volume of 5 μL of water to a 0.5 mL tube.
2. Denature sample in a thermal cycler for 10 min at 70 °C.
3. Assemble polyadenylation master mix using the online calculator. Gently vortex and centrifuge.
4. Add 5 μL of the polyadenylation master mix to each sample.
5. Incubate at 37 °C for 15 min. During this incubation, you may want to prepare the first-strand synthesis master mix (see below).
6. Remove the samples from the incubator, place on ice, and proceed immediately to the next step.

3.6.2 First-Strand Synthesis

7. Prepare the first-strand master mix using the online calculator. Gently vortex and centrifuge.

8. Add 10 μL of the master mix to each sample. Gently vortex and centrifuge.
9. Incubate for 2 h at 42 °C. Then place on ice, and proceed immediately to the second-strand cDNA synthesis.

3.6.3 Second-Strand Synthesis

10. On ice, assemble the second-strand master mix using the online calculator. Gently vortex and centrifuge.
11. Add 80 μL of the master mix to each sample. Gently vortex and centrifuge.
12. Incubate in a thermal cycler precooled to 16 °C for 2 h (the lid temperature should either match or be turned off). During this incubation, bring the bottle of nuclease-free water to 50 °C.
13. When the incubation is finished, place the samples on ice and proceed to the cDNA cleanup.

3.6.4 cDNA Cleanup

14. Add 250 μL of cDNA binding buffer and transfer to a cDNA cleanup spin cartridge.
15. Centrifuge for 1 min. Discard the flow through.
16. Pipette 500 μL of wash buffer onto the cartridge. Centrifuge for 1 min. Discard flow through.
17. Centrifuge for an additional minute to remove any trace amounts of ethanol.
18. Transfer the cartridge to a clean cDNA elution tube.
19. Elute by adding 18 μL of preheated 50 °C nuclease-free water to the cartridge. Incubate at room temperature for 2 min. Centrifuge for 1 min.
20. Discard the cartridge, and place the samples on ice.

3.6.5 In Vitro Transcription

21. Prepare the in vitro transcription master mix using the online calculator. Gently vortex and centrifuge.
22. Add 24 μL of the master mix to each sample. Gently vortex and centrifuge.
23. Incubate at 37 °C. A 14-h incubation time is recommended to maximize aRNA yield.
24. Add 60 μL of nuclease-free water to bring the final volume up to 100 μL and place on ice.

3.6.6 aRNA Purification

25. At least 30 min before starting the purification incubate the nuclease-free water at 55 °C.
26. Add 350 μL of aRNA binding buffer to each sample.
27. Add 250 μL of 100 % ethanol. Mix by pipetting up and down.
28. Transfer the mixture to an aRNA filter column. Centrifuge for 1 min. Discard flow through.

29. Apply 650 μL of wash buffer to the column. Centrifuge for 1 min. Discard flow through.
30. Centrifuge for an additional 2 min to remove any trace amounts of ethanol.
31. Transfer the cartridge to a clean collection tube.
32. Add 200 μL of the preheated 55 °C water to the center of the column. Place the column in incubator set at 55 °C for 10 min.
33. Centrifuge for 1.5 min. Discard the flow through. There should now be ~200 μL of purified aRNA.
34. Quantify using either a Nanodrop spectrophotometer or a RiboGreen-based fluorescence detection. This is a potential stopping point. Store at −80 °C.

3.7 cDNA Synthesis

Single-stranded RNA (or aRNA) is converted to cDNA via the Universal RiboClone cDNA Synthesis System using random primers. We typically use 10 μg of aRNA in the cDNA synthesis to obtain a final mass of ~5–8 μg cDNA. The amount of RNA used can be varied depending on the particular requirements for the sequencing platform. The reagent volumes depend on the amount of input RNA used and will have to be adjusted from those given in the manual. For example, the kit protocol is based on a 2 μg input RNA volume, but for many metatranscriptomic experiments a 10 μg amount will be desired, and this requires scaling the reagents in each step up by 5×. The appropriate volumes can be found in the Universal RiboClone cDNA Synthesis System manual.

3.7.1 First-Strand Synthesis

1. The input RNA sample for the first-strand synthesis should contain 10 μg aRNA in a volume of 65 μL. If the volume is <65 μL, bring up to 65 μL with nuclease-free water. If the aRNA volume is >65 μL, concentrate via either speed vacuum or ethanol precipitation.
2. Add random primers to the aRNA. Gently mix and centrifuge.
3. In a preheated thermal cycler, denature the RNA–primer mixture at 70 °C for 10 min. Immediately after, place the tubes on ice for 5 min.
4. Transfer the mixture to a 1.5 mL tube. This size is necessary to account for the increase in volume in the coming steps.
5. Add first-strand 5× buffer and RNasin ribonuclease inhibitor. Gently mix the reaction and briefly centrifuge.
6. Incubate the mixture in an incubator at 37 °C for 5 min.
7. Add sodium pyrophosphate, AMV reverse transcriptase, and nuclease-free water. Gently mix and centrifuge.
8. Incubate the mixture in incubator at 37 °C for 1 h. Afterward, place on ice and proceed directly to second-strand synthesis (*see* **Note 8**).

3.7.2 Second-Strand Synthesis

9. On ice, add the following components to the first-strand reaction: second-strand 5× buffer, BSA, DNA polymerase, RNase H, and nuclease-free water. Gently mix and centrifuge.
10. Incubate at 14 °C for 2 h (*see* **Note 9**).
11. Remove the second-strand reaction from the incubator or the thermal cycler, and add T4 DNA polymerase.
12. Return to the incubator or the thermal cycler set at 14 °C and incubate for another 10 min. The temperature in this step deviates from the kit protocol.
13. Add 10 μL of 0.1 mM EDTA per μg input RNA to stop the reaction, and place the mixture on ice.

3.7.3 QiaQuick Cleanup

The volumes below are based on a cDNA synthesis of 10 μg, for which the final volume in the second-strand synthesis is 550 μL.

14. To increase elution efficiency and reduce guanidinium thiocyanate carryover, warm the PE buffer to 37 °C for at least 2 h before using.
15. Divide the sample into two 275 μL aliquots placed in 2 mL tubes.
16. Add 688 μL PB buffer and mix thoroughly by vortexing (*see* **Note 10**).
17. Place a mini column on the vacuum manifold, and start the vacuum. Pipette the mixture from both tubes onto the column until the entire volume has passed through. Remove suction.
18. Remove the PE buffer from the 37 °C incubator, and add 750 μL PE to the column. Restore the vacuum until the buffer has passed through. Repeat the wash with another 750 μL PE.
19. Remove the cartridge from the manifold and place in a collection tube. Centrifuge at 9,400 ×*g* for 2 min to remove any residual wash solution. Transfer to a clean 1.5 mL tube.
20. Add 50 μL of nuclease-free water or TE buffer (*see* **Note 11**) and let stand at room temperature for 1 min. Centrifuge at 9,400 ×*g* for 2 min. Discard the cartridge. The cDNA is now ready for sequencing.

3.8 Bioinformatic Analysis

Processing of the resulting sequence reads involves quality trimming, internal standard quantification, residual rRNA identification and removal, and finally functional annotation of the protein-encoding reads. Several of these processing steps can be carried out using platforms freely available through CAMERA (http://camera.calit2.net) or MG-RAST (http://metagenomics.anl.gov). For more information *see* Bragg and Tyson, Chapter 15.

Next-generation reads can produce both systematic and random sequencing errors specific to the platform used. A quality metric (such as Phred) should be used to identify and remove low-quality regions or entire sequences. The number of internal standards recovered by the sequencing should be quantified by a BLASTn homology search and removed from further processing. Inevitably, some rRNAs will escape the rRNA reduction process and need to be removed to prevent misleading functional annotations of these sequences. A BLASTn homology search against the SILVA large and small rRNA subunit database (www.arb-silva.de) can be used to identify bacterial, archaeal, and eukaryotic rRNA sequences. Once identified, these sequences should be removed from further processing. Finally, potential protein-encoding reads can be annotated based on homology to databases that span a wide range of functional resolution from broad functions (COGs) to strain specific proteins (RefSeq). The calculations for total transcript pool size and individual transcript abundance (*see* [14] for more details) are calculated as follows:

$$P_a = \frac{P_s \times S_a}{S_s} \qquad T_a = \frac{T_s \times P_a}{P_s}$$

P_a = estimated total transcripts in the environmental sample, P_s = potential protein-encoding sequence reads (total number of sequence reads − rRNA sequences − S_s), S_a = molecules of internal standard added to the sample, S_s = internal standard sequence reads observed in the data set, T_s = reads assigned to a transcript category observed in the data set, and T_a = molecules of any particular transcript category in the sample. This can then be divided by the mass or the volume of sample collected to calculate the transcript abundance on a per environmental unit basis.

4 Notes

1. RNAs have short half-lives and are quickly degraded by ubiquitous RNAses. Wearing gloves, working in a clean lab space such as a PCR hood with a UV lamp, and being sure to clean all pipettes and surfaces with an RNase-degrading solution (e.g.: RNaseZap, Ambion) improve success rates. When not actively working with the RNA sample, it should be kept either on ice or frozen at −80 °C. Only plasticware that has been certified as nuclease free should be used. ART barrier tips are recommended for pipetting.
2. We recommend Supor filters (Pall, Port Washington, NY), although any 0.22 μm filter should be appropriate.
3. The transcriptional state of cells held in the collection apparatus for longer than 30 min may start to deviate significantly from the cells' in situ state.
4. Candidate internal standard sequences should be unique with respect to the environmental metatranscriptome. This can be

tested by comparing the proposed standard to relevant databases to identify any regions of homology that could interfere with unambiguous identification of the added standard in the sequence library. Multiple standards of different length and sequence composition should be designed. The addition of multiple standards to the sample helps to control for pipetting errors and size selection biases that may decrease the accuracy of the final quantification estimate.

5. Many filters are brittle when frozen and can be easily shattered with a mallet. Alternatively, the filters can be cut up with sterilized scissors.
6. Depending on how much biomass was on the original filter, it may take several minutes to pass all of the lysate through the column. (If a vacuum manifold is unavailable, the lysate can be passed through the spin column using multiple centrifugations).
7. It is also worthwhile to run the PCR products on a gel to confirm the correct product amplification and concentration.
8. Note that the first-strand synthesis with AMV reverse transcriptase may be done at 42 °C if oligo dT primers, rather than random hexamers, are used.
9. We have found it easiest to conduct the second-strand synthesis in a refrigerated incubator (cooled to 14 °C for at least an hour before using), as it does not require splitting up a single sample into multiple 0.5 mL tubes. However, if a reliable refrigerated incubator cannot be found, a thermal cycler can be used.
10. The mixture should be yellow. If orange or violet, the pH is not correct and will need to be adjusted (see kit manual).
11. During the cleanup, the cDNA can be eluted in either nuclease-free water or TE depending on the downstream requirements of sequencing.

Acknowledgements

We wish to thank Christa Smith for help in protocol development and Camille English for creating the figure artwork.

References

1. Poretsky RS, Bano N, Buchan A, LeCleir G, Kleikemper J, Pickering M et al (2005) Analysis of microbial gene transcripts in environmental samples. Appl Environ Microbiol 71:4121–4126
2. Poretsky RS, Hewson I, Sun SL, Allen AE, Zehr JP, Moran MA (2009) Comparative day/night metatranscriptomic analysis of microbial communities in the North Pacific subtropical gyre. Environ Microbiol 11:1358–1375

3. Poretsky RS, Sun S, Mou X, Moran MA (2010) Transporter genes expressed by coastal bacterioplankton in response to dissolved organic carbon. Environ Microbiol 12:616–627
4. Frias-Lopez J, Shi Y, Tyson GW, Coleman ML, Schuster SC, Chisholm SW et al (2008) Microbial community gene expression in ocean surface waters. Proc Natl Acad Sci U S A 105:3805–3810
5. Gilbert JA, Field D, Huang Y, Edwards R, Li W, Gilna P et al (2008) Detection of large numbers of novel sequences in the metatranscriptomes of complex marine microbial communities. PLoS ONE 3:e3042
6. Urich TA, Lanzen J, Qi DH, Huson DH, Schleper C, Schuster SC (2008) Simultaneous assessment of soil microbial community structure and function through analysis of the meta-transcriptome. PLoS ONE 3:e2527
7. Shi YM, Tyson GW, DeLong EF (2009) Metatranscriptomics reveals unique microbial small RNAs in the ocean's water column. Nature 459:266–269
8. Hewson I, Poretsky RS, Dyhrman ST, Zielinski B, White AE, Tripp HJ et al (2009) Microbial community gene expression within colonies of the diazotroph, *Trichodesmium*, from the Southwest Pacific Ocean. ISME J 3:1286–1300
9. Gifford SM, Sharma S, Rinta-Kanto JM, Moran MA (2011) Quantitative analysis of a deeply-sequenced marine microbial metatranscriptome. ISME J 5:461–472
10. He S, Wurtzel O, Singh K, Froula JL, Yilmaz S, Tringe SG, Wang Z et al (2010) Validation of two ribosomal RNA removal methods for microbial metatranscriptomics. Nat Methods 7:807–812
11. Stewart FJ, Ottesen EA, DeLong EF (2010) Development and quantitative analyses of a universal rRNA-subtraction protocol for microbial metatranscriptomics. ISME J 4: 896–907
12. Rivers A, Sharma S, Tringe SG, Martin J, Joye SB, Moran MA (2013) Transcriptional response of bathypelagic marine bacterioplankton to the deepwater horizon oil spill. ISME J. doi:10.1038/ismej.2013.129
13. Moran MA, Satinsky B, Gifford SM, Luo H, Rivers A, Chan LK et al (2013) Sizing up metatranscriptomics. ISME J 7(2):237–243
14. Satinsky BM, Gifford SM, Crump B, Moran MA (2013) Use of internal standards for quantitative metatranscriptome and metagenome analysis. Methods in enzymology: microbial metagenomics, metatranscriptomics, and metaproteomics. Elsevier, Amsterdam

Chapter 18

Quantitative Metaproteomics: Functional Insights into Microbial Communities

Chongle Pan and Jillian F. Banfield

Abstract

Quantitative metaproteomics aims to accurately determine the relative abundances of thousands of proteins in a microbial community. This approach can be used to provide a comprehensive view of metabolic activities of organisms in microbial communities and uncover significant changes in protein expression between communities at different developmental stages, environment types or in response to different perturbations. Here, we describe three strategies for quantitative metaproteomics, including label-free, ^{15}N metabolic labeling, and isobaric chemical labeling. The measurements are all based on a shotgun proteomics workflow involving proteolysis, two-dimensional liquid chromatogram-tandem mass spectrometry, and database searching against a metagenomic protein database. Quantitative metaproteomics was established and demonstrated using a model microbial community from acid mine drainage.

Key words Quantitative proteomics, Microbial community, Label-free, ^{15}N metabolic labeling, Isobaric chemical labeling, Liquid chromatography, Tandem mass spectrometry

1 Introduction

A microbial community is made up of many organisms at different abundance levels. Each of these organisms maintains a complement of proteins, or a proteome, to carry out a variety of metabolic activities [1, 2]. Because it is difficult to physically separate different organisms, all community cells in a field-collected sample are lysed directly and their proteomes are extracted together as a complex protein mixture. A community proteome sample is generally measured using a shotgun approach that starts with the digestion of all proteins into peptides [3]. Trypsin, the enzyme typically used for the digestion, specifically cleaves proteins in the C-terminus of arginine and lysine residues. The peptide mixture is then analyzed using two-dimensional liquid chromatography–tandem mass spectrometry (LC-MS/MS) [4]. An automated 24-h LC-MS/MS run typically produces millions of MS/MS scans that record the fragmentation patterns of many isolated peptides.

Ian T. Paulsen and Andrew J. Holmes (eds.), *Environmental Microbiology: Methods and Protocols*, Methods in Molecular Biology, vol. 1096, DOI 10.1007/978-1-62703-712-9_18, © Springer Science+Business Media, LLC 2014

In data analysis, a database searching computer program is used to identify peptides in a sample by matching each MS/MS scan against the predicted fragmentation pattern of every peptide in a protein sequence database [5]. Identified peptides are assigned to proteins from different organisms in the community. Single amino acid polymorphisms that distinguish homologous proteins often allow resolution of proteins from different strains in a community [6]. A protein cannot be identified if its peptide sequences are not represented in the sequence database. Thus, metaproteomics requires high-quality metagenomic sequences from the same or a highly similar community. While some protein identifications can be achieved using genome sequences of related isolate microorganisms, missing species- and strain-specific candidate sequences will lower protein identification rates and potentially introduce biases that will confound abundance estimates.

Microbial communities can respond to environmental perturbations by adjusting protein expression and organismal composition. Measurement of protein abundance changes in a treatment sample relative to a control sample provides valuable insights into this behavior. Here we describe three approaches to quantitative metaproteomics that determines the abundance changes of thousands of proteins in community proteomes. In the label-free approach, different field-collected samples from a microbial community are prepared identically in parallel and analyzed in separate LC-MS/MS runs [7]. Relative abundances of proteins are estimated using spectral counts, which are the numbers of MS/MS scans matched to peptides and proteins identified in a run [8]. The label-free approach has been widely used to study a variety of microbial communities [9]. Although simple to implement, it has a limited quantification accuracy and precision because of the technical variability between separate sample preparations and LC-MS/MS runs.

The ^{15}N metabolic labeling approach requires culturing of a microbial community in a defined ^{15}N-enriched medium to label all nitrogen with ^{15}N in every protein [10]. This provides a ^{15}N-labeled reference sample to compare an unlabeled field-collected sample against. Samples of the reference and field-collected sample are mixed in equal amounts. Every protein has two mass-different isotopic variants: an unlabeled isotopologue and a heavier labeled isotopologue. The abundance ratios between the two isotopic variants of peptides can be determined from the intensities of their mass spectrometric signals, which are then used to estimate relative abundances of proteins. Mixing the samples prior to protein preparation and LC-MS/MS analysis avoids potential biases that could otherwise be introduced if samples were processed and measured separately. This approach was used to study the pH perturbation and ecological succession of an acid mine drainage community [11, 12].

The isobaric chemical labeling approach is based on two similar commercial kits: Isobaric Tag for Relative and Absolute Quantification (iTRAQ) [13] and Tandem Mass Tag (TMT) [14]. Protein extraction and trypsin digestion are performed separately for each field-collected sample under comparison. Peptides from different samples are then labeled with different isobaric variants of an iTRAQ tag or a TMT tag. The peptide samples are pooled together for the LC-MS/MS analysis. Relative abundances of peptides are measured based on the intensities of reporter ions in the MS/MS scans. The isobaric chemical labeling approach removes technical variability in LC-MS/MS analysis, but not in sample preparation. 6-plex TMT and 8-plex iTRAQ allow comparing up to six and eight samples in a single LC-MS/MS analysis, respectively.

Here we describe the experimental and computational procedures for the three quantitative metaproteomics approaches. Protocols are provided for the shotgun proteomics measurement for all three approaches. The sample preparation is based on the FASP technology [15]. The LC-MS/MS analysis is based on the MudPIT technology [16] with split-phase back column [17]. This provides a starting point for optimizing measurements for specific microbial communities.

2 Materials

2.1 Proteome Sample Preparation

1. SDS cell lysis buffer: 5 % SDS, 50 mM dithiothreitol (DTT), 50 mM Tris–HCl, pH 8.
2. Protein resuspension buffer: 6 M guanidine, 50 mM DTT.
3. TEAB buffer: 100 mM triethyl ammonium bicarbonate (TEAB).
4. IAA solution: 50 mM iodoacetamide (IAA), 100 mM TEAB.
5. Filter Aided Sample Preparation (FASP) protein digestion kit (Expedeon, San Diego, CA).
6. Sequencing-grade modified trypsin (Promega, Madison, WI).
7. Bicinchoninic acid (BCA) protein assay kit.
8. TMT kit (Thermo Scientific, San Jose, CA) or iTRAQ kit (AB SCIEX, Framingham, MA) (*see* **Note 1**).

2.2 LC-MS/MS Analysis

1. High-performance liquid chromatography (HPLC) with solvent A (95 % H_2O, 5 % CH_3CN, and 0.1 % HCOOH), solvent B (30 % H_2O, 70 % CH_3CN, and 0.1 % HCOOH), and solvent D (500 mM CH_3COONH_4 in solvent A).
2. Nanospray ionization source (Proxeon, Cambridge MA).
3. LTQ Orbitrap or Exactive Orbitrap mass spectrometer (Thermo Scientific, San Jose, CA) (*see* **Note 2**).
4. PicoFrit nanospray column (New Objective, Woburn, MA).

2.3 Data Analysis

1. Raxport program for FT1/FT2 input file generation, freely available at Raxport.omicsbio.org.
2. Sipros program [18–20] for protein identification, freely available from Sipros.omicsbio.org.
3. ProRata program [21, 22] for protein quantification, freely available from ProRata.omicsbio.org.
4. Metagenomic protein database in FASTA format.

3 Methods

3.1 Experimental Design

Select an appropriate quantitative proteomics approach based on the following guidelines [23]:

1. The label-free approach (LF) simply measures each community sample under comparison separately. The measurement results are combined during data analysis to derive relative quantification information using spectral counting. This requires a reproducible sample preparation workflow to reduce sample-to-sample variability and a stable LC-MS/MS platform to decrease run-to-run variability. The label-free approach typically cannot determine small abundance changes or accurately measure abundance changes.
2. Metabolic labeling (ML) requires growth of a ^{15}N-labeled reference microbial community from a small amount of unlabeled inoculum using a ^{15}N-enriched growth medium. The reference community should have a similar organism membership as unlabeled communities collected from the field. Not all natural communities can be cultivated in the laboratory, so this may be a limitation of the ML approach. However, if this can be accomplished, multiple field-collected samples can be indirectly compared to each other by using a common reference sample. After mixing the samples under comparison, technical variability of protein quantification is avoided in all downstream steps.
3. Isobaric chemical labeling (ICL) uses TMT or iTRAQ to label peptides from different samples and mix them together for LC-MS/MS analysis. The detection of low-molecular-weight reporter ions requires an Orbitrap or a time-of-flight mass spectrometer. 6-plex TMT and 8-plex iTRAQ allows multiplexing up to six and eight samples to significantly increase the number of samples measured in an instrument day.
4. Both ML and ICL provide accurate and precise estimation of fold changes of many proteins between multiple samples. ML introduces less variability in protein preparation, but requires growing a representative sample of the microbial

community in a ^{15}N-enriched growth medium (which may not be possible). ICL allows multiplexing many samples, but requires specific types of mass spectrometers. LF is most straightforward to perform, but offers lower quantification performance.

3.2 Proteome Sample Preparation

1. Collect the samples. For ML, mix equal aliquot of each unlabeled field-collected sample with the ^{15}N-labeled reference sample (*see* **Note 3**) and process the combined samples in all subsequent steps. For ICL, process each field-collected sample separately in parallel until chemical labeling in **step 10** and combine labeled samples for LC-MS/MS. For LF, process and measure each field sample separately and combine the results in data analysis.
2. Add the SDS cell lysis buffer to the field samples and incubate at 99 °C for 15 min (*see* **Note 4**). Centrifuge at 6,000 × *g* for 20 min at 4 °C to remove debris. Collect the supernatants as the whole-cell lysates (*see* **Note 5**).
3. Add 100 % TCA to the whole-cell lysates at the ratio of 1:4 by solution volume. Precipitate proteins at 4 °C overnight. Centrifuge at 15,000 × *g* for 10 min at 4 °C and remove supernatants. Wash the protein pellets three times with chilled acetone. Dry the protein pellets in room temperature.
4. Dissolve the protein pellets in the protein resuspension buffer and incubate at 60 °C for 1 h (*see* **Note 6**). Determine the total protein concentrations of the protein samples using the BCA assay.
5. Load the protein samples containing 50–200 μg of proteins onto the FASP spin filters (*see* **Note 7**). Centrifuge at 14,000 × *g* for 15 min and discard the flow-through (*see* **Note 8**).
6. Wash the protein samples on filter: add 200 μL TEAB buffer to the FASP spin filters, centrifuge at 14,000 × *g* for 15 min, and discard the flow-through. Repeat the wash.
7. Add 100 μL IAA solution to the spin filters. Vortex and incubate for 20 min in the dark (*see* **Note 9**). Centrifuge at 14,000 × *g* for 15 min to remove the IAA solution. Wash the protein samples on filter with 200 μL TEAB buffer twice.
8. Dissolve trypsin in 75 μL TEAB buffer and add to the spin filters. Add 2 μg trypsin per 100 μg of protein. Incubate at 37 °C overnight.
9. Transfer the spin filters to new tubes to collect the filtrates as peptide samples (*see* **Note 10**). Add 50 μL TEAB buffer and centrifuge the spin filters at 14,000 × *g* for 15 min. Repeat this twice.

10. Determine the peptide concentration of the peptide samples using a BCA assay. For LF or ML, the peptide samples are ready for LC-MS/MS analysis. For CIL, perform iTRAQ or TMT labeling on the peptide samples following the manufacturer's protocol (*see* **Note 11**) and combine the labeled peptide samples for LC-MS/MS analysis.

3.3 Liquid Chromatography–Tandem Mass Spectrometry

1. Pack a split-phase back column with 3 cm of C18 reverse phase (RP) resin (Luna, Phenomenex) and 3 cm of strong cation exchange (SCX) resin (Luna, Phenomenex) (*see* **Note 12**). Pack a front column in the PicoFrit nanospray column with 15 cm of C18 RP resin (*see* **Note 13**).
2. Load 10–50 μg of peptides onto the back column using a pressure cell. De-salt the sample for 15 min with 100 % solvent A. Elute peptides from the RP phase to the SCX phase of the back column using a 1-h gradient from 100 % solvent A to 100 % solvent B.
3. Connect the back column to the front column. Start the two-demensional LC separation consisting of 12 cycles. Each cycle starts with a short salt pulse of solvent D to elute an aliquot of peptides from the SCX phase of the back column onto the RP phase of the front column. The peptides are then eluted using a continuous 2-h RP gradient and directly electrosprayed into the mass spectrometer. The 12 cycles use increasing percentages of solvent D in the salt pulses to create a step gradient for the SCX separation.
4. Ionized peptides are measured by LTQ Orbitrap. Use the following parameters for LF: MS1 scans using Orbitrap at resolution 30,000 with 1-microscan averaging and 20 data-dependent MS2 scans with collisional induced dissociation (CID) using LTQ (*see* **Note 14**). Use the following parameters for ML: MS1 scans using Orbitrap at resolution 30,000 with 2-microscan averaging and ten data-dependent MS2 scans with CID using LTQ (*see* **Note 15**). Use the following parameters for CIL: MS1 scans using Orbitrap at resolution 30,000 with 1-microscan averaging and ten data-dependent MS2 scans with higher-energy collisional induced dissociation (HCD) using Orbitrap at resolution 7,500 with 1-microscan averaging (*see* **Note 16**).

3.4 Data Processing

1. Convert the acquired MS/MS data from the RAW format to the FT1/FT2 format using the Raxport program.
2. Search the MS/MS data against the metagenomic protein database using the Sipros program (*see* **Note 17**). Select an appropriate Sipros configuration file based on the type of analysis. The ML data is searched in two iterations, one for unlabeled peptides using normal amino acid masses and the other

for ^{15}N-labeled peptides using ^{15}N-labeled amino acid masses. The CIL data is searched with dynamic modification of peptides by the chemical tags (*see* **Note 18**).

3. Filter the search results at 1 % false discovery rate (FDR) estimated at the peptide level (*see* **Note 19**). Assemble peptide identifications into protein identifications with a minimum of two peptides per protein and one unique peptide per protein.
4. Perform quantification calculation using the ProRata program (*see* **Note 20**). Relative abundances of proteins are estimated with normalized spectral counts for LF analysis, relative intensities of peptides' chromatographic peaks for ML analysis, and relative intensities of peptides' reporter ions for CIL analysis. Estimate the statistical significance of abundance changes from replicate measurements by using Student's *t*-tests or rank product tests with multiple comparison correction. Identify proteins with significant changes based on their *p*-values and relative abundances.

4 Notes

1. TMT and iTRAQ are very similar in terms of experimental protocol, cost per sample, and quantification performance. Up to two, four, six, and eight samples can be compared in a single LC-MS/MS run using duplex TMT, 4-plex iTRAQ, 6-plex TMT, and 8-plex iTRAQ, respectively.
2. Both instruments can acquire high-resolution MS1 and MS2 in Orbitrap. The more expensive LTQ Orbitrap can also acquire low-resolution MS2 in LTQ with collisional induced dissociation.
3. Equal amounts of biomass should be mixed for ML. If it is difficult to estimate the amount of biomass in field samples, process the samples to **step 4** and mix equal amounts of total protein extract.
4. The cell lysis method should be optimized for the microbial community under study to achieve unbiased cell lysis for all organisms and efficient protein desorption from the sample matrix. Additional sample treatments, such as sonication, liquid nitrogen grinding, and bead beating, can be used to assist the cell lysis.
5. SDS is detrimental to the LC-MS/MS analysis. The high concentration of SDS in the whole-cell lysates is removed by the subsequent TCA precipitation and FASP.
6. The protein resuspension buffer contains 6 M guanidine to help dissolving protein pellets and 50 mM DTT to reduce disulfide bonds in proteins.

7. The FASP procedure described below is similar to the manufacturer's protocol. The key modification is to use the TEAB buffer, instead of a urea solution, to be compatible with ICL.
8. Denatured proteins are retained on top of the molecular-weight cutoff filter (typically 30 kDa cut-off) in the FASP spin filter. Low-molecular-weight chemicals pass through the filter with the solution. The filter can retain many small proteins less than 30 kDa, probably because of the unfolded form of denatured proteins. As particulates in the sample may clog the filter, it is important to remove particulates in the previous steps.
9. IAA alkylation blocks free cysteine residues from forming disulfide bonds.
10. Trypsin digests proteins to peptides. Peptides can pass through the filter.
11. Because iTRAQ and TMT react with primary amines, the labeling requires a buffer that does not contain any reagent with primary amines. The TEAB buffer of the peptide samples is compatible with iTRAQ and TMT labeling. A vial of the labeling reagent can label up to 100 μg of peptides. Adding too much peptides may result in incomplete labeling and inaccurate quantification.
12. A new back column should be used for every LC-MS/MS run to minimize sample carry-over between runs. Pack the back column as described [17].
13. A front column can be reused for multiple runs with extensive wash after each run. The PicoFrit column is integrated with the electrospray tip to eliminate post-column dead-volume.
14. Quantification in the LF approach is based on spectral counting. This setting maximizes the MS/MS acquisition rate.
15. Quantification in the ML approach is based on selected ion chromatograms reconstructed from MS1 scans. This setting uses 2-microscans averaging to obtain high-quality MS1 scans and only ten MS2 scans to acquire more MS1 scans per minute.
16. Quantification in the CIL approach is based on reporter ions in the MS2 scans. HCD and Orbitrap MS2 detection is needed to accurately measure reporter ions with low mass-to-charge ratios.
17. Database searching is a computing-intensive task that requires a high-end computer with multi-core CPUs. Sipros can also use large computer clusters.
18. Searches with dynamic modifications can identify peptides with and without chemical modification. Too many peptides identified without chemical modification indicate incomplete labeling reaction.

19. FDR is the percentage of false identifications out of all identification. Reverse sequences are typically used as decoy sequences for estimating false-discovery rate (FDR) of identification [24]. FDR is calculated as the percentage of decoy identifications out of all identifications.
20. Nonunique peptides can be attributed to multiple proteins in the database. Many peptides from closely related species or strains are nonunique. Quantification may be biased by including nonunique peptides in protein relative abundance estimation. Use only unique peptides for quantification if microorganisms in the community have large sequence divergence from one another.

Acknowledgement

We like to thank Zhou Li, Nathan C. VerBerkmoes, Robert L. Hettich, Christopher P. Belnap, and Nagiza F. Samatova for their contributions in the development of the methodology described here. This work was funded by the US Department of Energy, Office of Biological and Environmental Research, and Office of Advanced Scientific Computing Research. Oak Ridge National Laboratory is managed by UT-Battelle LLC for the Department of Energy.

References

1. Tyson GW, Chapman J, Hugenholtz P, Allen EE, Ram RJ, Richardson PM, Solovyev VV, Rubin EM, Rokhsar DS, Banfield JF (2004) Community structure and metabolism through reconstruction of microbial genomes from the environment. Nature 428(6978): 37–43
2. Ram RJ, Verberkmoes NC, Thelen MP, Tyson GW, Baker BJ, Blake RC 2nd, Shah M, Hettich RL, Banfield JF (2005) Community proteomics of a natural microbial biofilm. Science 308(5730):1915–1920
3. Hettich RL, Pan C, Chourey K, Giannone RJ (2013) Metaproteomics: harnessing the power of high performance mass spectrometry to identify the suite of proteins that control metabolic activities in microbial communities. Anal Chem 85(9):4203–4214
4. Link AJ, Eng J, Schieltz DM, Carmack E, Mize GJ, Morris DR, Garvik BM, Yates JR 3rd (1999) Direct analysis of protein complexes using mass spectrometry. Nat Biotechnol 17(7):676–682
5. Eng JK, Mccormack AL, Yates JR (1994) An approach to correlate tandem mass-spectral data of peptides with amino-acid-sequences in a protein database. J Am Soc Mass Spectrom 5(11):976–989
6. Lo I, Denef VJ, Verberkmoes NC, Shah MB, Goltsman D, DiBartolo G, Tyson GW, Allen EE, Ram RJ, Detter JC, Richardson P, Thelen MP, Hettich RL, Banfield JF (2007) Strain-resolved community proteomics reveals recombining genomes of acidophilic bacteria. Nature 446(7135):537–541
7. Justice NB, Pan C, Mueller R, Spaulding SE, Shah V, Sun CL, Yelton AP, Miller CS, Thomas BC, Shah M, VerBerkmoes N, Hettich R, Banfield JF (2012) Heterotrophic archaea contribute to carbon cycling in low-pH, suboxic biofilm communities. Appl Environ Microbiol 78(23):8321–8330
8. Liu H, Sadygov RG, Yates JR 3rd (2004) A model for random sampling and estimation of relative protein abundance in shotgun proteomics. Anal Chem 76(14):4193–4201
9. VerBerkmoes NC, Denef VJ, Hettich RL, Banfield JF (2009) Systems biology: functional analysis of natural microbial consortia using community proteomics. Nat Rev Microbiol 7(3):196–205
10. Belnap CP, Pan C, Verberkmoes NC, Power ME, Samatova NF, Carver RL, Hettich RL, Banfield JF (2010) Cultivation and quantitative proteomic analyses of acidophilic microbial communities. ISME J 4:520–530

11. Belnap CP, Pan C, Denef VJ, Samatova NF, Hettich RL, Banfield JF (2011) Quantitative proteomic analyses of the response of acidophilic microbial communities to different pH conditions. ISME J 5(7):1152–1161
12. Mueller RS, Dill BD, Pan C, Belnap CP, Thomas BC, VerBerkmoes NC, Hettich RL, Banfield JF (2011) Proteome changes in the initial bacterial colonist during ecological succession in an acid mine drainage biofilm community. Environ Microbiol 13(8): 2279–2292
13. Ross PL, Huang YN, Marchese JN, Williamson B, Parker K, Hattan S, Khainovski N, Pillai S, Dey S, Daniels S, Purkayastha S, Juhasz P, Martin S, Bartlet-Jones M, He F, Jacobson A, Pappin DJ (2004) Multiplexed protein quantitation in Saccharomyces cerevisiae using amine-reactive isobaric tagging reagents. Mol Cell Proteomics 3(12):1154–1169
14. Thompson A, Schafer J, Kuhn K, Kienle S, Schwarz J, Schmidt G, Neumann T, Johnstone R, Mohammed AK, Hamon C (2003) Tandem mass tags: a novel quantification strategy for comparative analysis of complex protein mixtures by MS/MS. Anal Chem 75(8): 1895–1904
15. Wisniewski JR, Zougman A, Nagaraj N, Mann M (2009) Universal sample preparation method for proteome analysis. Nat Methods 6(5):359–362
16. Washburn MP, Wolters D, Yates JR 3rd (2001) Large-scale analysis of the yeast proteome by multidimensional protein identification technology. Nat Biotechnol 19(3):242–247
17. McDonald WH, Ohi R, Miyamoto DT, Mitchison TJ, Yates JR (2002) Comparison of three directly coupled HPLC MS/MS strategies for identification of proteins from complex mixtures: single-dimension LC-MS/MS, 2-phase MudPIT, and 3-phase MudPIT. Int J Mass Spectrom 219(1):245–251
18. Pan C, Fischer CR, Hyatt D, Bowen BP, Hettich RL, Banfield JF (2011) Quantitative tracking of isotope flows in proteomes of microbial communities. Mol Cell Proteomics 10(4):M110.006049
19. Hyatt D, Pan C (2012) Exhaustive database searching for amino acid mutations in proteomes. Bioinformatics 28(14):1895–1901
20. Wang Y, Ahn TH, Li Z, Pan C (2013) Sipros/ProRata: a versatile informatics system for quantitative community proteomics. Bioinformatics 29(16):2064–2065
21. Pan C, Kora G, McDonald WH, Tabb DL, VerBerkmoes NC, Hurst GB, Pelletier DA, Samatova NF, Hettich RL (2006) ProRata: a quantitative proteomics program for accurate protein abundance ratio estimation with confidence interval evaluation. Anal Chem 78(20):7121–7131
22. Pan C, Kora G, Tabb DL, Pelletier DA, McDonald WH, Hurst GB, Hettich RL, Samatova NF (2006) Robust estimation of peptide abundance ratios and rigorous scoring of their variability and bias in quantitative shotgun proteomics. Anal Chem 78(20): 7110–7120
23. Li Z, Adams RM, Chourey K, Hurst GB, Hettich RL, Pan C (2012) Systematic comparison of label-free, metabolic labeling, and isobaric chemical labeling for quantitative proteomics on LTQ Orbitrap Velos. J Proteome Res 11(3):1582–1590
24. Peng J, Elias JE, Thoreen CC, Licklider LJ, Gygi SP (2003) Evaluation of multidimensional chromatography coupled with tandem mass spectrometry (LC/LC-MS/MS) for large-scale protein analysis: the yeast proteome. J Proteome Res 2(1):43–50

INDEX

Ian T. Paulsen and Andrew J. Holmes (eds.), *Environmental Microbiology: Methods and Protocols*, Methods in Molecular Biology, vol. 1096, DOI 10.1007/978-1-62703-712-9,

MIX
Papier aus verantwortungsvollen Quellen
Paper from responsible sources
FSC® C105338

If you have any concerns about our products,
you can contact us on
ProductSafety@springernature.com

In case Publisher is established outside the EU,
the EU authorized representative is:
Springer Nature Customer Service Center GmbH
Europaplatz 3, 69115 Heidelberg, Germany

Printed by Libri Plureos GmbH
in Hamburg, Germany